Seismic Design of Building Structures

A Professional's Introduction
to Earthquake Forces
and Design Details

Sixth Edition

Michael R. Lindeburg, PE

Professional Publications, Inc.
Belmont, CA

SEISMIC DESIGN OF BUILDING STRUCTURES
Sixth Edition

Printed in the United States of America

Professional Publications, Inc.
1250 Fifth Avenue, Belmont, CA 94002
(415) 593-9119

Current printing of this edition: 1

Library of Congress Cataloging-in-Publication Data
Lindeburg, Michael R.
    Seismic design of building structures : a professional's
introduction to earthquake forces and design details / Michael R.
Lindeburg. -- 6th ed.
        p.   cm.
    Includes indexes.
    ISBN 0-912045-76-0
    1. Earthquake engineering--California--Problems, exercises, etc.
    2. Civil engineering--California.   I. Title.
TA654.6.L56   1994
624.1' 762' 09794--dc20   94-8517
                                                    CIP

TABLE OF CONTENTS

PROFESSIONAL PUBLICATIONS, INC. ● Belmont, CA

PART 5:   RESPONSE OF STRUCTURES

PART 6:   SEISMIC PROVISIONS OF
THE UNIFORM BUILDING
CODE

PART 7:   DIAPHRAGM THEORY

PROFESSIONAL PUBLICATIONS, INC. • Belmont, CA

PART 8: GENERAL STRUCTURAL DESIGN

PART 9: DETAILS OF SEISMIC-RESISTANT CONCRETE STRUCTURES

PART 10: DETAILS OF SEISMIC-RESISTANT STEEL STRUCTURES

PART 11: DETAILS OF SEISMIC-RESISTANT MASONRY STRUCTURES

PART 12: DETAILS OF SEISMIC-RESISTANT WOOD STRUCTURES

PREFACE TO THE SIXTH EDITION

This sixth edition has been updated to conform to the 1991 edition of the *Uniform Building Code* (UBC) and the 1990 SEAOC *Lateral Force Requirements* ("Blue Book"). For the purpose of most engineers using this book (i.e., studying for their civil engineering licensing exams), the seismic provisions of the 1991 UBC do not differ significantly from those of the 1988 UBC. However, there are some subtle differences, and the two codes differ markedly in numbering.

In addition to bringing the text into conformance with the 1991 UBC and revising the book's code references, I have written extensive commentary, added 35 new solved questions and two new appendices, and made many minor clarifications, reorganizations, and corrections. Many of these changes were suggested by readers and review course instructors who took the time to tell me how to make the book more useful.

To help you cross-reference with the UBC, I have provided a third index (to accompany the indexes for subjects and tables) listed by UBC section. It is not a very exciting new feature, but I think you will find it useful.

Since the seismic examination was originally implemented, the California Board of Registration for Professional Engineers has received numerous comments, suggestions, and complaints from nonstructural examinees. These examinees feel (and rightfully so) that the depth of structural analysis experience and code-related knowledge expected of them far exceeds anything they will ever be faced with in practice.

In response to these comments, there has been a lot of discussion about shifting the exam emphasis away from highly-technical structural and calculation-intensive questions and toward conceptual and general familiarity questions. Thus far, the shift has been mostly a matter of discussion, not of action. Regardless, this book is intended to prepare you adequately for both types of questions.

After you take your examination, I will be indebted to you for your comments on this book. A postage-paid comment card appears in the back for that purpose. I always acknowledge readers' suggestions, and I appreciate any suggestions you think will help future engineers.

Michael R. Lindeburg, P.E.
Belmont, CA
March, 1994

PREFACE TO THE FIFTH EDITION

This fifth edition of *Seismic Design* is a concise introduction to basic seismic concepts and principles and is particularly intended for individuals taking the engineering (P.E.) and architectural (A.R.E.) licensing exams. I want to emphasize the descriptive word *introduction*, however, as I have consistently summarized subjects that entire books have been written about.

Since my goal in writing this new edition was to make the subject intuitive for you, this book is somewhat different from the previous four editions. For example, there is much less emphasis on vibration theory in this edition than in the previous edition. Unfortunately, the trend in seismic problems on licensing examinations has been toward practical structural design and away from theory and commentary, so this edition also emphasizes more code provisions and is more rigorous. It has been my intention to do this without sacrificing the simple explanations needed to develop an intuitive feeling toward the subject.

Treatment of all subjects has been greatly expanded in this edition. To make it possible to "jump into" a subject without first reading preparatory sections, many cross-references to material in other sections and to actual UBC code sections have been included. In keeping pace with changes in seismic knowledge and technology, this edition includes detailed material from the new UBC seismic code first introduced in 1988, flexible diaphragms, connections, fasteners, and other construction details. The older seismic code, introduced in the 1980 *Blue Book* and last used in the 1985 UBC, is not mentioned or covered at all in this book. The older UBC is well covered by the numerous seismic design books currently in print. Many are listed in this book's bibliography.

In addition to numerous new examples, the format of the practice problems in this edition has changed. Relatively complete solutions to all practice problems are included in this book, and these solutions follow immediately after the problem statement. In addition, this book contains a detailed index, set of definitions, and complete listing of nomenclature.

As I was writing this text, I recognized that there were exceptions to almost every general statement I made, particularly those statements that simplify the theory sufficiently to make the subject intuitive (which was my intention). In order to make this book as complete and rigorous as possible without complicating things excessively, I explained most of these exceptions in footnotes.

None of the analytical techniques required in earthquake problems are obscure or proprietary. However, the nomenclature and terminology is often unique to the seismic community. This book, with its many examples and explanations, presents these techniques in step-by-step format. The explanations draw upon practical information, code requirements, and simple vibrational theory. Complex proofs and derivations have been omitted.

As I previously stated, this book is written with the full expectation that it will be used as a study guide and reference for solving seismic design and analysis problems on professional engineering and architectural licensing examinations. Of course, this book has other uses, as readers of all my books regularly report. I hope this book will take its place in your library as a reference that you use throughout your career.

Michael R. Lindeburg, P.E.
August 1990
Belmont, CA

PROFESSIONAL PUBLICATIONS, INC. • Belmont, CA

ACKNOWLEDGMENTS

There is nothing in this publication representing original research, and I am indebted to the many professors, authors, scientists, and engineers that have contributed to my knowledge of seismology and seismic design techniques.

I am lucky to have had the early manuscripts reviewed by several outstanding specialists in the earthquake field. I am indebted to Vitelmo V. Bertero, Director of the Earthquake Engineering Research Center at the University of California (Berkeley, CA), John G. Shipp, P.E., Senior Technical Manager of EQE Incorporated (Costa Mesa, CA), Albert Tung, Ph.D., P.E., of Stanford's Civil Engineering Department (Stanford, CA), and James E. Onderka, P.E. (Orinda, CA), all of whom reviewed portions of the manuscript. Reading their comments was sometimes a humbling experience, but considering their improvements, I wouldn't have wanted it any other way.

Also, Karie Youngdahl helped with another Professional Publications book by checking my punctuation, grammar, and spelling.

Of course, the material presented in this publication is ultimately my responsibility. And, while I hope there are none, any errors are attributable to me and me alone.

I am very grateful to the International Conference of Building Officials (ICBO) for its generous permission to reproduce so many of the tables found in the Uniform Building Code (UBC). These tables are critical to understanding and using the code-related provisions of seismic design. The value of this publication has been greatly enhanced by ICBO's generosity.

Other publishers also cooperated in giving permission to reproduce important tables and figures. They are credited where their material appears.

Acknowledgment is made that many of the practice problems at the end of this book are based on or derived from problems that appeared in previous years' California engineering licensing examinations. These problems are used with permission of the California Board of Professional Engineers and Land Surveyors, which has copyrighted the original material.

I cannot forget to mention some of the individuals at Professional Publications that turned my scribbles and scratchings into a real book. After they worked at warp speed to produce a new edition of my *Engineer-In-Training Reference Manual*, they agreed, for some unknown reason, to take this publication from manuscript to book form in only three months (a near impossibility in the publishing world). Joanne Bergeson managed the overall project; Wendy Nelson conceived the cover design; Lisa Rominger designed the book and oversaw all production functions; she, Mary Christensson, and Sylvia Osias typeset the text; Yves Martin produced the illustrations, and Lynelle Dodge provided artwork for the cover. Kurt Stephan was the primary proofreader of the typesetting and illustrations.

Finally, I acknowledge the unwavering support of my family which, as usual, has had to put up with my habitual writing. My wife, Elizabeth, and my two daughters, Jenny and Katie, lost the family time that went into this book. They may be beginning to understand how I think and what is important to me. If not, they never seem to complain.

NOMENCLATURE

Unless defined otherwise in the text, the following symbols are used in this book. Consistent units are presented. In some cases (such as modulus of elasticity and drift), however, it is customary to report values in smaller units (lbf/in^2 and in).

Symbol	Term	U.S.	SI
a	acceleration	ft/sec^2	m/s^2
a	link beam distance	ft	m
A	amplitude	ft	m
A	area	ft^2	m^2
b	link beam distance	ft	m
b	parallel wall length	ft	m
B	damping coefficient	lbf-sec/ft	N·s/m
B	seismic parameter	–	–
C	chord force	lbf	N
C	numerical coefficient	–	–
C	seismic parameter	mi^{-2}	km^{-2}
C_t	numerical coefficient	–	–
d	depth	ft	m
D	column depth	ft	m
D	dead load	lbf	N
D	fault slip	ft	m
e	eccentricity	ft	m
E	unfactored earthquake load	lbf	N
E	modulus of elasticity	lbf/ft^2	Pa
E	energy released	ft-lbf	J
f	frequency	Hz	Hz
F	force	lbf	N
F	story shear	lbf	N
g	acceleration of gravity	ft/sec^2	m/s^2
g_c	gravitational constant (32.2)	ft-lbm/ lbf-sec²	n.a.
G	shear modulus	lbf/ft^2	Pa
h	height	ft	m
h	story height	ft	m
h_n	height above base to level n	ft	m
H	horizontal force	lbf	N
H	story height	ft	m
I	importance factor	–	–
I	moment of inertia	ft^4	m^4
J	polar moment of inertia	ft^4	m^4
k	stiffness (spring constant)	lbf/ft	N/m
L	length	ft	m

Symbol	Term	U.S.	SI
L	live load	lbf	N
m	mass	slug	kg
m	mass (customary U.S.)	lbm	n.a.
M	moment	ft-lbf	N·m
M	Richter magnitude	–	–
n	cycle number	–	–
n	exponent	–	–
N	number of earthquakes	–	–
P	sum of dead and live loads	lbf	N
P	magnitude of forcing function	lbf	N
PGA	peak ground acceleration	ft/sec^2	m/s^2
r	radius or moment arm	ft	m
r	resistance	ohms	ohms
R	relative rigidity	–	–
R_w	numerical coefficient	–	–
s	distance	ft	m
S	site/soil coefficient	–	–
S_a	spectral acceleration	ft/sec^2	m/s^2
S_d	spectral displacement	ft	m
S_v	spectral velocity	ft/sec	m/s
t	thickness	ft	m
t	time	sec	s
T	fundamental period of vibration	sec	s
U	strain energy	$ft\text{-}lbf/ft^3$	J/m^3
U	ultimate capacity	lbf	N
v	shear stress	lbf/ft^2	Pa
v	velocity	ft/sec	m/s
V	base shear	lbf	N
w	load per unit length	lbf/ft	N/m
W	weight	lbf	N
x	drift	ft	m
x	position or excursion	ft	m
y	height over which wind acts	ft	m
Y	number of years	–	–
Z	seismic zone factor	–	–

Symbol	Term	Units	
		U.S.	SI
β	magnification factor	–	–
Γ	participation factor	–	–
Δ	drift	ft	m
ϵ	strain	–	–
θ	stability coefficient	–	–
λ	Lame's constant	–	–
μ	ductility factor	–	–
ξ	damping ratio	–	–
ρ	rock/soil density	lbm/ft^3	kg/m^3
σ	stress	lbf/ft^2	Pa
τ	short period of time	sec	s
ϕ	mode shape factor	–	–
ω	angular frequency	rad/sec	rad/s

Subscripts

0	initial or calibration
d	damped
i	inertial or level number
j	mode number
n	cycle number
p	part or portion
P	P-wave
R	resilience
st	static
sx	between levels $x-1$ and x
S	S-wave
t	top or total
T	toughness
x	level number
x	with respect to x-axis
y	with respect to y-axis

PROFESSIONAL PUBLICATIONS, INC. ● Belmont, CA

1
BASIC SEISMOLOGY

1 THE NATURE OF EARTHQUAKES

An earthquake is an oscillatory, sometimes violent movement of the Earth's surface that follows a release of energy in the Earth's crust. This energy can be generated by a sudden dislocation of segments of the crust, a volcanic eruption, or a man-made explosion. Most of the destructive earthquakes, however, are caused by dislocations of the crust.

When subjected to geologic forces from plate tectonics, the crust initially strains (i.e., bends and shears) elastically. For pure axial loading, Hooke's law gives the stress that accompanies this strain.

$$\sigma = E\epsilon \quad \text{[axial loading]} \qquad [1]$$

As rock is stressed, it stores *strain energy*, U. The elastic strain energy per unit volume for pure axial loading is[1]

$$U = \frac{\sigma\epsilon}{2} \quad \text{[axial loading]} \qquad [2]$$

When the stress exceeds the ultimate strength of the rocks, the rocks break and quickly move (i.e, they "snap") into new positions. In the process of breaking, the strain energy is released and *seismic waves* are generated. This is the basic description of the *elastic rebound theory* of earthquake generation.[2]

---

[1]When the rock is stressed in shear, an analogous term for shear strain energy can be written. Both energy forms can be present simultaneously.

[2]The *dilatational source theory* explains that earthquakes are produced from the explosion, sudden vaporization, or implosion of underground material. However, this theory is no longer favored as a description of the source of earthquakes. Other sources, such as Wiegel (1970), cover this theory in greater detail.

These waves travel from the source of the earthquake (known as the *hypocenter* or *focus*) to more distant locations along the surface of and through the Earth. (The wave velocities depend on the nature of the waves and the material through which the waves travel. See Sec. 14.) Some of the vibrations are of high enough frequency to be audible, while others are of very low frequency with periods of many seconds and thus are inaudible.

A new theory may explain how some earthquakes are triggered. Geologists know that pumping fluids into the ground under high pressure can trigger earthquakes. Now there is evidence from the gas-producing regions in France that removing fluids from pores deep in the earth can also trigger earthquakes. Oil and gas are the main fluids of concern; pumping water from aquifers close to the surface is probably not as likely to result in an earthquake. The theory states that the reservoir shrinks when the gas or oil is removed, but the rocks surrounding the reservoir do not. This results in stresses in the earth that later are released in an earthquake.

2 EARTHQUAKE TERMINOLOGY

The *epicenter* of an earthquake is the point on the Earth's surface directly above the *focus* (also known as the *hypocenter*). The location of an earthquake is commonly described by the geographic position of its epicenter and its focal depth. The *focal depth* of an earthquake is the depth from the Earth's surface to the focus. These terms are illustrated in Fig. 1.

Earthquakes with focal depths of less than approximately 60 km (40 mi) are classified as *shallow earthquakes*. Very shallow earthquakes are caused by the fracturing of brittle rock in the crust or by internal strain energy that overcomes the friction locking opposite sides of a fault. California earthquakes are typically shallow.[3] *Intermediate earthquakes*, whose causes are not fully understood, have focal depths ranging from 60 to 300 km (40 to 190 mi). *Deep earthquakes* may have focal depths of up to 700 km (450 mi).

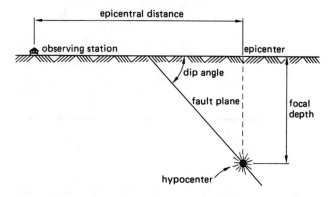

Figure 1   Earthquake Terminology

The slip propagates from the epicenter along the fault with a velocity up to that of the outward-radiating seismic shear wave front—about 3 km/s (1.8 mi/sec)—until the entire affected segment is in motion. (See Sec. 14 for a description of shear waves.)

3 GLOBAL SEISMICITY

Most earthquakes occur in areas bordering the Pacific Ocean.[4] This circum-Pacific belt, nicknamed the *ring of fire*, includes the Pacific coasts of North America and South America, the Aleutian Islands, Japan, Southeast Asia, and Australia. The reason for such a concentration is explained by plate tectonics theory. (See Sec. 5.)

The United States has experienced less destruction than other countries located in this earthquake zone. This is partly due to the country's relatively young age and attention to earthquake-resistant construction methods, but millions of Americans now live in potential earthquake areas. Large parts of the western United States are known to be vulnerable to damage from earthquakes.[5] Nowhere is this more true than in California.

---

[3]Since there is no deep subduction zone, earthquakes in California typically occur at depths of less than 15 km (10 mi).

[4]The other major concentration of earthquakes is in a much smaller east-west belt that runs between Asia and the Mediterranean.

[5]It is interesting that the largest earthquakes on the North American continent in the history of the United States occurred in the east (the 1811 and 1812 New Madrid, Missouri and the 1886 Charleston, South Carolina earthquakes, the latter of which had an estimated Richter magnitude in excess of 8.2). However, earthquakes in these regions are much less frequent than earthquakes in the western United States.

Nuclear reactors, dams, schools, hospitals, and high-rise buildings are planned and built in locations of high seismic hazard. This has created an urgent need for greater attention to the mitigation of earthquake-induced damage.

4 CONTINENTAL DRIFT

It has been known since the early 1900s that the continents are moving relative to one another, movement known as *continental drift*. In fact, fossilized records of past climates (the subject of the field of *paleoclimatology*) indicate that the continents have been moving slowly about the globe for millions of years. For example, the same 300-million-year-old fossilized deposits are found in India and in the Arctic.

The theory of continental drift was reasonably established during the 1930s but was not universally accepted. In the 1950s, the emerging science of *paleomagnetism* provided new supporting evidence of continental drift. Many rocks, such as volcanic rock solidified from molten lava, contain tiny grains of magnetic minerals such as *magnetite*. When these minerals are formed, they retain the magnetic orientation of the Earth's magnetic field at the time of their formation. The magnetic orientations of rocks suggest the same ancient locations of the continents suggested by paleoclimatology and other geologic criteria.

An enormous amount of geophysical data was gathered during the 1950s and 1960s, particularly from oceanographic research vessels such as the *Omar Challenger*. A system of interconnecting submarine ridges, called *mid-ocean ridges*, was discovered circling the Earth. Such ridges are located approximately midway between continents that are moving apart (e.g., between Africa and South America). It is now recognized that new oceanic crust is being formed at the ridges and is added to the plates moving apart. This is known as *seafloor spreading*.

Great submarine trenches were also located, particularly along the convex oceanic sides of the volcanic arcs that make up the Pacific ring of fire. Inclined zones of earthquakes dip down from these trenches to as deep as 700 km (450 mi) into the mantle beneath and behind the volcanic arcs. Oceanic crust is formed at spreading ridges behind the moving plates.

The ocean crust, known to consist of alternating belts of highly and weakly magnetic oceanic crust material, represents magnetic records as new crust forms in the gaps behind separating plates. The symmetrical belts record the ambient magnetism on opposite sides of the newly-formed ridges.

The crust is destroyed at the same rate elsewhere as oceanic plates dip down at the trenches and slide deep into the mantle along the seismic zones. However, there is a global balance between crust formation and destruction. The formation of plate material in the Atlantic Ocean is compensated by absorption of plate material, primarily in the Pacific Ocean.

5 PLATE TECTONICS

Most earthquakes are a manifestation of the fragmentation of the Earth's outer shell (known as the *lithosphere*) into various large and small plates. (The academic field that studies plate motion is known as *plate tectonics*.) There are seven very large plates, each consisting of both oceanic and continental portions. There are also a dozen or more small plates, not all of which are shown in Fig. 2.

Each plate is approximately 80 to 100 km (50 to 60 mi) thick and has thick and thin parts. The thinner part deforms by elastic bending and brittle breakage. The thicker part yields plastically. Beneath the plate is a viscous layer on which the entire plate slides. The plates themselves tend to be internally rigid, interacting only at the edges.

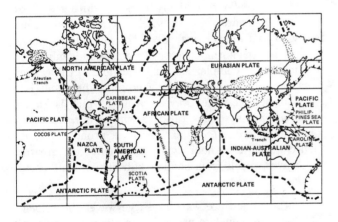

Figure 2 Lithosphere Plates

These plates move relative to each other with steady velocities that approach 0.13 m/a (0.4 ft/yr).[6] Although plate velocities are slow by human standards, they are extremely rapid geologically. For example, a motion of 0.05 m/a (0.15 ft/yr) adds up to 50 km (30 mi) in only 1 million years. Some plate motions have been continuous for 100 million years.

Depending on location, the plates can be moving apart, colliding slowly to build mountain ranges, or slipping laterally past or sliding over and under one another.

6 SUBMARINE RIDGES

Where plates are pulling apart, particularly along the system of submarine ridges, hot material from the deeper mantle wells up to fill the gap. Some of the mantle material appears as lava in volcanic material. Most solidifies beneath the surface, forming a submarine ridge. The ridge is high relative to the ocean bottom because the mantle material is hot and, hence, low in density.

As the plates move apart, the ridge material gradually cools and contracts, and its surface sinks. Ridges generally form step-like alterations in height perpendicular to the direction of plate motion. Strike-slip faults form parallel to the direction of plate motion. (See Sec. 10.)

7 SUBMARINE TRENCHES

Where plates converge, one dips down and slides beneath the other in a process known as *subduction*. Generally, an oceanic plates slides, or subducts, beneath a continental plate (as is happening along the west coast of South America) or beneath another oceanic plate (as is happening along the east side of the Philippine Sea plate). A trench is formed where the subducting plate dips down. The sediment from the ocean floor is scraped off against the front edge of the top plate. This is illustrated in Fig. 3.[7]

Far back under the top plate, inclined zones of earthquakes reach down into the mantle. The average depth of these zones is approximately 125 km (80 mi), but the zones can approach 700 km (450 mi) in depth. The hypocenters of earthquakes in these zones indicate the trajectory of the subducted plate.

A belt of volcanoes typically occurs above this earthquake zone, roughly paralleling the plate edges. (The Pacific Ocean coastal region of South America is typical of such an area.) Rock melting, which ultimately produces the volcanoes, starts when water combined in the crystalline structures of various minerals, or otherwise trapped, is removed by the increase in pressure on the subducted plate. The water loss lowers the net energy required to melt the remaining rock.

---

[6]There are grounds for suggesting that the African plate may be fixed relative to the deep mantle. If so, it is the only major plate that is fixed.

[7]Details and dimensions are those for western Java and the Java trench system, but other systems are similar.

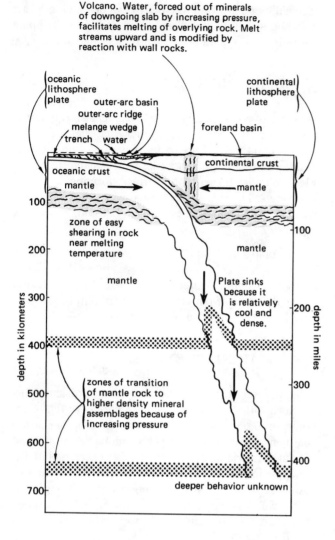

Volcano. Water, forced out of minerals of downgoing slab by increasing pressure, facilitates melting of overlying rock. Melt streams upward and is modified by reaction with wall rocks.

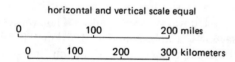

horizontal and vertical scale equal

Figure 3   Zone of Subduction

8 EARTHQUAKE ENERGY RELEASE[8]

Shallow earthquakes represent sudden slippages and are accompanied by a release of elastic strain energy stored in the rock over a long period. It is not totally clear whether deep mantle subduction-zone earthquakes are accompanied by similar elastic releases (i.e., the elastic rebound theory) or are merely abrupt contractions of part of the subducting plate into rock of higher density

(i.e., an aspect of the dilatational theory). However, recent research seems to indicate the former explanation is more appropriate than the latter.

Only a fraction of the energy released in an earthquake actually appears in seismic waves. Most of the released strain energy is reabsorbed locally by the moving, deforming, and heating of the rock. The fraction absorbed increases irregularly with increasing size of earthquakes. Minor earthquakes generally do not represent a sufficient release of energy to dissipate the strain energy and prevent great earthquakes, although a slow creep along a fault can provide a partial release. A great earthquake, however, does not necessarily release all of the strain energy either.

Great earthquakes occur primarily along convergent (subducting) plate boundaries.[9] Submerged ridges (where plates are spreading apart) are so hot at relatively shallow depths that the solid rock above them cannot store enough elastic strain energy to produce great earthquakes. The infrequent large earthquakes that do occur in these ridge systems are mostly on the longer strike-slip faults. (See Sec. 10.)

9 SEISMIC SEA WAVES

When the seafloor suddenly rises up during a great earthquake, water also rises with it and then rushes away to find a level surface. If the floor drops, water rushes in. An enormous mass of water is suddenly set in motion, and a complex sloshing back and forth between continents continues for many hours. The result is a train of surface-water waves, each of which is known as a *seismic sea wave* (also known as a *tidal wave*), or, in Japanese, a *tsunami*. The most pronounced sudden changes in seafloor depth, and hence the greatest sea waves, result from shallow subduction-zone earthquakes.

As with any surface wave or surge wave, the velocity of a tsunami depends primarily on the ocean depth.[10] In deep ocean, waves travel at about 800 km/h (500 mi/hr). The waves at sea may be an hour apart and perhaps only 0.3 m (1 ft) in height. Combined with a wave period of 5 to 60 minutes, they are virtually undetectable. As a wave approaches land, however, the wave velocity decreases due to increased friction with

[8]See Sec. 20 also.

[9]The great 1906 San Francisco earthquake, however, was not a subduction-zone earthquake.

[10]This is greatly simplified. There has been much research on the effect of depth and other aspects of tsunami generation and propagation. A good survey of the subject is contained in Murty (1977).

the increasingly shallow seafloor. As the wave velocity decreases, the wave height increases.

Where seafloor topography and orientation are optimal for tsunami formation (where there is a gently sloping seafloor and where the slope is parallel to wave direction), the wave can form a wall of water more than 15 m (50 ft) in height. Such a wave can cause enormous destruction when it rushes onto shore. Nearby coastal points, where the bottom configuration is much different (i.e., more abrupt in depth change), may see the same wave pass as only a rapid surge and withdrawal of water.

Only normal (dip-slip) and thrust (reverse) faults produce tsunamis. The greater the depth of water, the larger the energy content of the tsunami.

10 FAULTS

A fault is a fracture in the Earth's crust along which two blocks have slipped relative to each other. One crustal block may move horizontally in one direction while the opposite block moves horizontally in the opposite direction. Alternatively, one block may move upward while the other moves downward.

One of the ways movement along faults can occur is by sudden displacement, or *slip*, of the crust or rock along a fault. During the 1906 San Francisco earthquake, the ground was displaced as much as 6.5 m (21 ft) in northern California along the San Andreas Fault. By comparison, the 1989 Loma Prieta earthquake had a maximum displacement of approximately 2 m (6 ft).

Most of the faults in California are vertical or near-vertical breaks. Movement along these breaks is predominantly horizontal in the northerly or northwesterly directions.[11] With *right-lateral movement* (such as the movement of earthquakes in the San Andreas system), a block on the opposite side of the fault (relative to an observer) moves to the right. Conversely, the block moves to the left in a *left-lateral fault*. Lateral movement is produced by *strike-slip (wrench) faults*.

A fault in which the movement is vertical is called a *dip-slip fault*. In a *normal fault*, the hanging wall moves down relative to the foot wall. In a *reverse fault*, also known as a *thrust fault*, the hanging wall moves up relative to the foot wall.

Along many faults movement is both horizontal and vertical. Such faults are named by combining the names

of each kind of movement. For example, Fig. 4 shows a left-lateral normal fault. The term *oblique fault* is also used.

A few *reverse faults* have been active in California. The planes of such faults are inclined to the Earth's surface. The rocks above the fault plane have been thrust upward and over the rocks below the fault plane. The Arvin-Tehachapi earthquake of 1952 was caused by the White Wolf reverse fault. The San Fernando earthquake of 1971 was caused by a sudden rupture along a reverse fault at the foot of the San Gabriel Mountains.

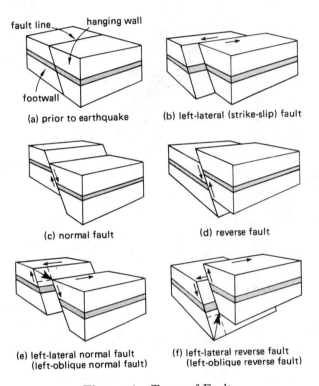

(a) prior to earthquake

(b) left-lateral (strike-slip) fault

(c) normal fault

(d) reverse fault

(e) left-lateral normal fault (left-oblique normal fault)

(f) left-lateral reverse fault (left-oblique reverse fault)

Figure 4   Types of Faults

11 CALIFORNIA FAULTS

The most earthquake-prone areas in the United States are those that are adjacent to the San Andreas Fault system of California, as well as the fault system that separates the Sierra Nevada from the Great Basin. Many of the individual faults of these major systems are known to have been active during the past 200 years. Others are believed to have been active since the end of the last great ice advance about 10,000 years ago.

---

[11]Notable exceptions are the Garlock and Big Pine left-lateral faults, which trend westerly.

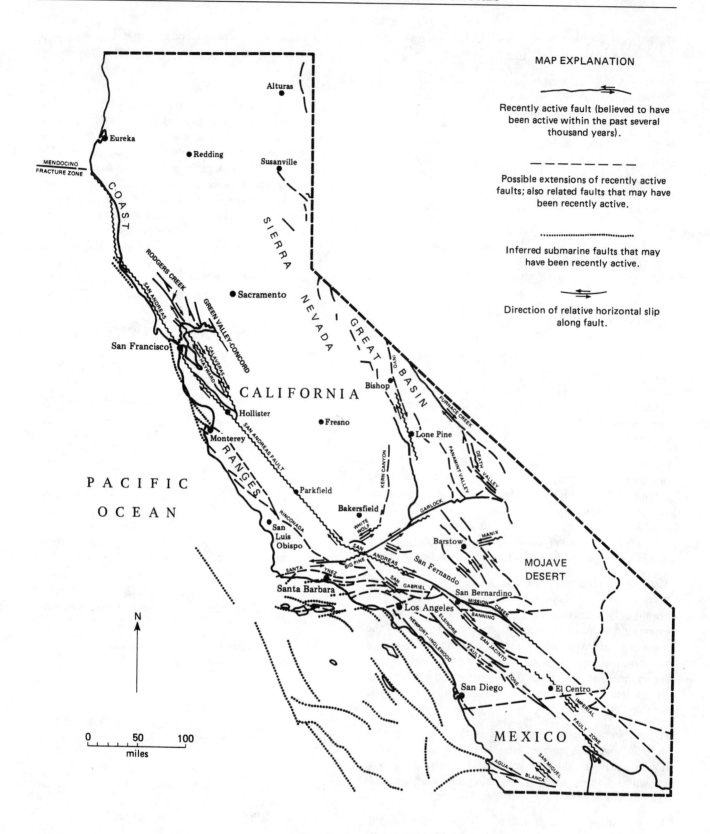

Figure 5   Active California Faults

During the past 200 years, many of the faults shown in Fig. 5 have experienced observed sudden slip or slow creep. Activity of other faults, however, can only be inferred from geologic and topographic relations that indicate the faults have been active during the past several thousand years. Such activity suggests that these faults could slip or creep again. (See Sec. 13.)

Earthquakes in California are relatively shallow and are clearly related to movement along active faults. Many California earthquakes have produced surface rupture.

12 SAN ANDREAS FAULT

The San Andreas Fault is the major fault of a network that cuts through rocks of the California coastal region. This right-lateral fault is a huge fracture more than 950 km (600 mi) long. It extends almost vertically into the Earth to a depth of at least 30 km (20 mi). In detail, it is a complex zone of crushed and broken rock from only a few feet wide to a mile wide. Many smaller faults branch from and join the San Andreas Fault.

A linear trough in the surface of the Earth reveals the presence of the San Andreas Fault over much of its length. From the air, the linear arrangements of lakes, bays, and valleys is apparent. On the ground, the fault zone can be recognized by long, straight escarpments, narrow ridges, and small, undrained ponds formed by the settling of small areas of rock. However, people on the ground usually do not realize when they are on or near the fault.

Geologists who have studied the fault between Los Angeles and San Francisco have suggested that the total accumulated displacement along the fault may be as much as 550 km (350 mi). Similarly, geological study of a segment of the fault between the Tejon Pass and the Salton Sea has revealed geologically similar terrains on opposite sides of the fault separated by 250 km (150 mi). This indicates that the separation is a result of movement along the San Andreas and branching San Gabriel faults.

Since 1934, earthquake activity along the San Andreas Fault system has been concentrated in three areas: (1) an off-shore area at the northernmost tip of the fault known as the *Mendocino fracture zone*, (2) the area along the fault between San Francisco and Parkfield, and (3) the southernmost fault section roughly bounded by Los Angeles and the border with Mexico. Creep, slip, and moderate earthquakes have occurred on a regular basis in these areas.

The two zones between these three active areas have had almost no earthquakes or known slip since the great earthquakes of 1857 in the southern segment and 1906 in the northern segment. This implies that these two zones of the San Andreas Fault system are temporarily locked and that strain energy is building. The lack of seismic activity in the locked sections could mean that the sections are subject to less frequent but larger fault movements and, correspondingly, more severe earthquakes.

13 CREEP

In addition to fault slip, a second type of fault movement known as *creep* can occur. Creep is characterized by continuous or intermittent movement without noticeable earthquakes. Fault creep occurring on portions of the Hayward, Calaveras, and San Andreas faults has produced cumulative offsets ranging from mere millimeters to almost 0.3 m (1 ft) in curbs, streets, and railroad tracks.

The offsets observed seem consistent with the creep rate measured. Precise surveying shows a slow drift at the average rate of approximately 5 cm/a (2 in/yr) along the San Andreas Fault. At that rate, over 550 km (350 mi) of offset has occurred during that past 100 million years.

14 SEISMIC WAVES

Seismic waves are of three types: compression, shear, and surface waves. Compression and shear waves travel from the epicenter through the Earth's interior to distant points on the surface. Only compression waves, however, can pass through the Earth's molten core. Because *compression waves* (also known as *longitudinal waves*) travel at great speeds (5800 m/s, or 19,000 ft/sec, in granite) and ordinarily reach the surface first, they are known as *P-waves* (for "primary waves").[12] P-wave velocity is given by Eq. 3.

$$v_P = \sqrt{\frac{\lambda + 2G}{\rho}} \qquad \text{[SI]} \qquad \text{[3(a)]}$$

$$v_P = \sqrt{\frac{(\lambda + 2G)g_c}{\rho}} \qquad \text{[U.S.]} \qquad \text{[3(b)]}$$

Shear waves (also known as *transverse waves*) do not travel as rapidly (3000 m/s, or 10,000 ft/sec, in granite) through the Earth's crust and mantle as do compression waves. Because they ordinarily reach the surface later,

---

[12]The term *dilatation* is used to describe negative compression (i.e., the "expansion" of rock from its normal density). (See Fig. 6.)

they are known as *S-waves* (for "secondary waves"). Instead of affecting material directly behind or ahead of their lines of travel, shear waves displace material at right angles to their path. Equation 4 gives the velocity of S-waves. While S-waves travel more slowly than P-waves, they transmit more energy and cause the majority of damage to structures.

$$v_S = \sqrt{\frac{G}{\rho}} \quad \text{[SI]} \qquad \text{[4(a)]}$$

$$v_S = \sqrt{\frac{G g_c}{\rho}} \quad \text{[U.S.]} \qquad \text{[4(b)]}$$

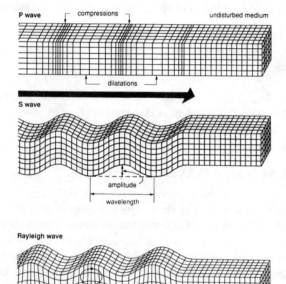

Figure 6   Types of Seismic Waves

Surface waves, also known as *R-waves* (for "Rayleigh waves") or *L-waves* (for "Love waves"), may or may not form. They arrive after the primary and secondary waves. In granite, R-waves move at approximately 2700 m/s (9000 ft/sec).

15 LOCATING THE EPICENTER

The first indication of an earthquake will often be a sharp "thud" signaling the arrival of the compression wave front. This will be followed by the shear waves and then the ground roll caused by the surface waves. The times separating the arrivals of the compression and shear waves at various seismometer stations can be used to locate the epicenter's position and depth.

The distance, *s*, from a seismometer to the epicenter can be determined from the wave velocities and the observed time between the arrival of the compression (P-) and shear (S-) waves.[13]

$$t_S - t_P = \left(\frac{1}{v_S} - \frac{1}{v_P} \right) s \qquad \text{[5]}$$

The epicenter and hypocenter correspond to the locations of initial fault slip but do not necessarily coincide with the center of energy release. For small and medium earthquakes (i.e., Richter magnitude $M < 6$), the points of initial fault slip and energy release are relatively close. For larger earthquakes, however, hundreds of kilometers can separate the two.

---

[13]It is not always possible to accurately determine the difference in arrival times from the seismometer record.

2

EARTHQUAKE CHARACTERISTICS

16 INTENSITY SCALE

The *intensity* of an earthquake (not to be confused with *magnitude*, see Sec. 18) is based on the damage and other observed effects on people, buildings, and other features. Intensity varies from place to place within the disturbed region. An earthquake in a densely populated area that results in many deaths and considerable damage may have the same magnitude as a shock in a remote area that does nothing more than frighten the wildlife. Large magnitude earthquakes that occur beneath the oceans may not even be felt by humans.

An intensity scale consists of a series of responses, such as people awakening, movement of furniture, and damage to chimneys. Although numerous intensity scales have been developed, the scale encountered most often in the United States is the *Modified Mercalli Intensity scale*, developed in 1931 by the American seismologists Harry Wood and Frank Neumann.[14]

The Modified Mercalli scale consists of 12 increasing levels of intensity (expressed as Roman numerals following the initials MM) that range from imperceptible shaking to catastrophic destruction. The lower numbers of the intensity scale generally are based on the manner in which the earthquake is felt by people. The higher numbers are based on observed structural damage. The numerals do not have a mathematical basis and therefore are more meaningful to nontechnical people than to those in technical fields.

---

[14]The original Mercalli scale was developed in 1902 by the Italian seismologist and volcanologist of the same name. The *Rossi-Forel scale* (its ten values are used in Fig. 7 to describe the 1906 San Francisco earthquake) was developed in the 1880s.

Table 1
Modified Mercalli Intensity Scale

Intensity	Observed Effects of Earthquake
I	Not felt except by very few under especially favorable conditions.
II	Felt only by a few persons at rest, especially by those on upper floors of buildings. Delicately suspended objects may swing.
III	Felt quite noticeably by persons indoors, especially in upper floors of buildings. Many people do not recognize it as an earthquake. Standing vehicles may rock slightly. Vibrations similar to the passing of a truck. Duration estimated.
IV	During the day, felt indoors by many, outdoors by a few. At night, some awakened. Dishes, windows, doors disturbed; walls make cracking sound. Sensation like heavy truck striking building. Standing vehicles rock noticeably.
V	Felt by nearly everyone; many awakened. Some dishes, windows broken. Unstable objects overturned. Pendulum clocks may stop.
VI	Felt by all, many frightened. Some heavy furniture moved. A few instances of fallen plaster. Damage slight.
VII	Damage negligible in buildings of good design and construction; slight to moderate in well-built ordinary structures; considerable damage in poorly-built structures. Some chimneys broken.

(continued on next page)

PROFESSIONAL PUBLICATIONS, INC. ● Belmont, CA

Table 1 (continued)

Intensity	Observed Effects of Earthquake
VIII	Damage slight in specially-designed structures; considerable damage in ordinary substantial buildings, with partial collapse. Damage great in poorly-built structures. Fallen chimneys, factory stacks, columns, monuments, walls. Heavy furniture overturned.
IX	Damage considerable in specially-designed structures; well-designed frame structures thrown out of plumb. Damage great in substantial buildings, with partial collapse. Buildings shifted off foundations.
X	Some well-built wooden structures destroyed; most masonry and frame structures with foundations destroyed. Rails bent.
XI	Few, if any, masonry structures remain standing. Bridges destroyed. Rails bent greatly.
XII	Damage total. Lines of sight and level are distorted. Objects thrown into air.

17 ISOSEISMAL MAPS

It is possible to compile a map of earthquake intensity over a region. Data for such an *isoseismal map* can be obtained by observation or, in some cases, by questionnaires mailed to residents after an earthquake.

18 RICHTER MAGNITUDE SCALE

In 1935, Charles F. Richter of the California Institute of Technology developed the Richter magnitude scale to measure earthquake strength. The magnitude, M, of an earthquake is determined from the logarithm of the amplitude recorded by a seismometer. (See Secs. 19 and 29.) Adjustments are included in the magnitude to compensate for the variation in the distance between the various seismometers and the epicenter. Because the Richter magnitude is a logarithmic scale, each whole number increase in magnitude represents a ten-fold increase in measured amplitude.

Richter magnitude is expressed in whole numbers and decimal fractions. For example, a magnitude of 5.3 might correspond to a moderate earthquake. A strong earthquake might be rated at 7.3. *Great earthquakes* have magnitudes above 7.5. Earthquakes with magnitudes of 2.0 or less are known as *microearthquakes*. While recorded on seismometers, microearthquakes are rarely felt by people.

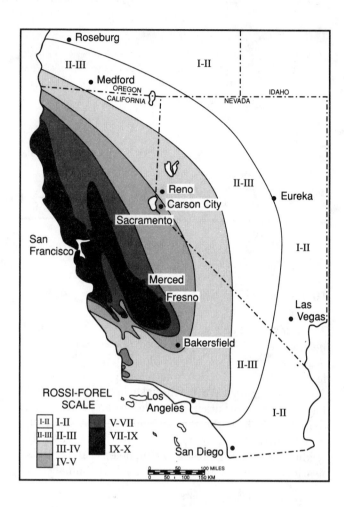

Figure 7 Isoseismal Map of 1906 San Francisco Earthquake (Based on the Rossi-Forel Intensity Scale)

Reprinted by permission of Prentice-Hall, Inc. and Carnegie Institution, from *Earthquake Engineering*, Robert L. Wiegel, ed., copyright © 1970 by Prentice-Hall, Inc. Redrawn from map 23 in *Report of the State Earthquake Investigation Committee*, Atla, Carnegie Institution of Washington, publication 87 (1908).

Several thousand seismic events with magnitudes of approximately 4.5 or greater occur each year and are strong enough to be recorded by seismometers all over the world. Earthquakes of this size and below have little potential to cause structural damage. Great earthquakes, such as the 1906 San Francisco earthquake and the 1964 Alaskan earthquake, occur, on the average, once each year.

The magnitude of an earthquake depends on the length and breadth of the *fault slip*, as well as on the amount of slip. The largest examples of fault slip recorded in California accompanied the earthquakes of 1857, 1872, and 1906—all of which had estimated magnitudes over 8.0 on the Richter scale.

Although the Richter scale has no lower or upper limit (i.e., it is "open ended"), the largest known shocks have had magnitudes in the 8.7 to 8.9 range.[15] The actual factor limiting energy release—and hence Richter magnitude—is the strength of the rocks in the Earth's crust.[16]

Because of the physical limitations of the faults and crust in the area, earthquakes larger than 8.5 in southern California are considered to be highly improbable.

19 RICHTER MAGNITUDE CALCULATION

The Richter magnitude, M, is calculated from the maximum amplitude, A, of the seismometer trace, as illustrated in Fig. 8. A_0 is the seismometer reading produced by an earthquake of standard size (i.e., a *calibration earthquake*). Generally, A_0 is 0.001 mm.

$$M = \log_{10}\left(\frac{A}{A_0}\right) \qquad [6]$$

Equation 6 assumes that a distance of 100 km (62 mi) separates the seismometer and the epicenter. For other distances, the nomograph of Fig. 9 and the following procedure can be used to calculate the magnitude. Due to the lack of reliable information on the nature of the Earth between the observation point and the earthquake epicenter, an error of 10 to 40 km (5 to 20 mi) in locating the epicenter is not unrealistic.

step 1: Determine the time between the arrival of the P- and S-waves.

step 2: Determine the maximum amplitude of oscillation.

step 3: Connect the arrival time difference on the left scale and the amplitude on the right scale with a straight line.

step 4: Read the Richter magnitude on the center scale.

step 5: Read the distance separating the seismometer and the epicenter from the left scale.

Whereas one seismometer can determine the approximate distance to the epicenter, it takes three seismometers to determine and verify the location of the epicenter.

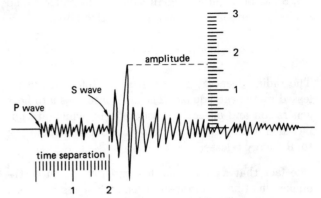

Figure 8 Typical Seismometer Amplitude Trace

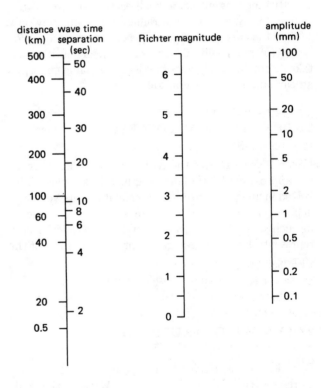

Figure 9 Richter Magnitude Correction Nomograph

---

[15]Depending on the calibration earthquake (see Sec. 19), earthquakes with negative Richter magnitudes can occur.

[16]It is said that Richter magnitudes much higher than the 8.7 to 8.9 range would correspond to an energy release sufficient to destroy the Earth itself. While this may be theoretically true for much larger Richter magnitudes (due to the logarithmic nature of the measurement), the limited strength of the rock itself serves to ensure that such a doomsday earthquake will never occur.

20 ENERGY RELEASE AND MAGNITUDE CORRELATION

Once the Richter magnitude, M, is known, an approximate relationship can be used to calculate the energy, E, radiated. Most of the relationships are of the form of Eq. 7.

$$\log_{10}E = \log_{10}E_0 + aM \qquad [7]$$

PROFESSIONAL PUBLICATIONS, INC. ● Belmont, CA

In 1956, Gutenberg and Richter determined the approximate correlation to be as given in Eq. 8. E is the energy in ergs. (See App. A for conversions to other units.) Although there have been other relationships developed, Eq. 8 has been verified against data from underground explosions and is the primary correlation cited.

$$\log_{10} E = 11.8 + 1.5M \qquad [8]$$

The radiated energy is less than the total energy released by the earthquake. The difference goes into heat generation and other nonelastic effects, which are not included in Eq. 8. Little is known about the amount of total energy release.

The fact that a fault zone has experienced an earthquake offers no assurance that enough stress has been relieved to prevent another earthquake. As indicated by the logarithmic relationship between seismic energy and Richter magnitude, a small earthquake (of magnitude 5, for example) would radiate approximately only 1/32 of the energy of an earthquake just one magnitude larger (of magnitude 6, for example). Thus it would take 32 small earthquakes to release the same energy as an earthquake one magnitude larger.

21 LENGTH OF ACTIVE FAULT

Equation 9 correlates the Richter magnitude, M, with the approximate total fault length, L in kilometers, involved in an earthquake. Such correlations are very site-dependent, and even then, there is considerable scatter in such data. Equation 9 should be considered only representative of the general (approximate) form of the correlation.

$$\log_{10} L = 1.02M - 5.77 \qquad [9]$$

22 LENGTH OF FAULT SLIP

Equation 10 (as derived by King and Knopoff in 1968) correlates the Richter magnitude, M, and the fault length, L (in meters), with the approximate length of vertical or horizontal fault slip, D (for *displacement*) in meters.[17] As with Eq. 9, this correlation should be considered representative of the general relationship.

$$\log_{10}(LD^2 \times 10^6) = 1.90M - 2.65 \qquad [10]$$

---

[17]King, Chi-Yu, and L. Knopoff, "Stress Drop in Earthquakes," *Bulletin of the Seismological Society of America*, **58** (1968): 249.

23 PEAK GROUND ACCELERATION

The *peak (maximum) ground acceleration*, PGA, is easily measured by a seismometer (see Sec. 29) or accelerometer (see Sec. 30) and is one of the most important characteristics of an earthquake.[18] PGA can be given in various units, including ft/sec^2, in/sec^2, or m/s^2. However, it is most common to specify PGA in "g's" (i.e., as a fraction or percent of gravitational acceleration).

$$\text{PGA} = \left(\frac{a_{\text{ft/sec}^2}}{32.2} \right) \times 100\% \qquad [11(a)]$$

$$= \left(\frac{a_{\text{in/sec}^2}}{386} \right) \times 100\% \qquad [11(b)]$$

$$= \left(\frac{a_{\text{m/s}^2}}{9.81} \right) \times 100\% \qquad [11(c)]$$

Significant ground accelerations in California include 1.25 g (Pacoima dam site, 1971 San Fernando earthquake), 0.50 g (1966 Parkfield earthquake), and 0.65 g (1989 Loma Prieta earthquake).

Equation 12 (as determined by Gutenberg and Richter in 1956) is one of many approximate relationships between the Richter magnitude, M, and the PGA at the epicenter. Of course, the ground acceleration (in rock) will decrease as the distance from the epicenter increases, and for this reason, equations of this type are called *attenuation equations*. (See Sec. 28.)

Attenuation equations are very site-dependent. Since Eq. 12 was developed, newer studies have resulted in better correlations in different formats and for many different locations, but they are based on limited data. Such studies are regularly incorporated into revisions of the seismic provisions of building codes.

$$\log_{10}\text{PGA} = -2.1 + 0.81M - 0.027M^2 \qquad [12]$$

Table 2 is a commonly cited correlation between magnitude, PGA, and duration of *strong-phase shaking* (see Sec. 29) in the vicinity of the epicenter of California earthquakes.[19] The values of acceleration in the table are somewhat on the high side. Ground acceleration in observed earthquakes usually has been lower.

---

[18]It is possible for an earthquake to exceed the range of the accelerometer or seismometer, in which case the PGA will not be recorded.

[19]This table gives the impression of high correlation even though the correlation is actually low. For example, the 1989 Loma Prieta earthquake magnitude was approximately 7.1, but the peak ground acceleration was 0.65 g.

Table 2
Approximate Peak Ground Acceleration and
Duration of Strong-Phase Shaking
(California Earthquakes)

Magnitude	Maximum Acceleration (g)	Duration (s)
5.0	0.09	2
5.5	0.15	6
6.0	0.22	12
6.5	0.29	18
7.0	0.37	24
7.5	0.45	30
8.0	0.50	34
8.5	0.50	37

Data from G. W. Housner, "Strong Ground Motion," in
Earthquake Engineering, Wiegel, ed., © 1970, p. 79. Reprinted
by permission of Prentice-Hall, Inc., Englewood Cliffs, NJ.

24 CORRELATION OF INTENSITY, MAGNITUDE, AND ACCELERATION

Although there are some empirical relationships, no exact correlations of intensity, magnitude, and acceleration are possible since many factors contribute to seismic behavior and structural performance. For example, seismic damage depends on the care that was taken at the time of building design and construction. Buildings in villages in undeveloped countries fare much worse than high-rise buildings in developed countries in earthquakes of equal magnitudes. This damage causes a corresponding lack of correlation between intensity and magnitude.

However, within a geographical region with consistent design and construction methods, fairly good correlation exists between structural performance and ground acceleration, because the Mercalli intensity scale is based specifically on observed damage.

Table 3
Approximate Relationship Between Mercalli Intensity
and Peak Ground Acceleration

MMI	PGA (g)
IV	0.03 and below
V	0.03–0.08
VI	0.08–0.15
VII	0.15–0.25
VIII	0.25–0.45
IX	0.45–0.60
X	0.60–0.80
XI	0.80–0.90
XII	0.90 and above

25 VERTICAL ACCELERATION

The shear (transverse) waves are at right angles to the compression (longitudinal) waves. (See Sec. 14.) Since there is nothing constraining the shear waves to a horizontal direction, it is not surprising that the S-wave (shear wave) can be broken down into horizontal and vertical components. When necessary, these are identified as SH-waves and SV-waves for horizontal and vertical shear waves, respectively.

Vertical ground acceleration is known to occur in almost all earthquakes. The peak vertical acceleration is usually approximately one-third of the peak horizontal acceleration, but often reaches a ratio of 2:3. Combined with resonance site effects, vertical forces can become substantial. Furthermore, forces from all three coordinate directions combine into a resultant force that can easily exceed the yield (and, sometimes, the ultimate) strength of a member.

The current UBC seismic design code is generally based on horizontal acceleration alone [Sec. 2334(a)]. This practice is justified by assuming that structures with horizontal seismic resistance will automatically have adequate vertical seismic resistance. One of the reasons this assumption has been accepted is that factors of safety should have been applied during the building design to ensure that a member is able to withstand a force equal to one gravity downward.

Experience has shown, however, that disregarding details to resist vertical forces can be a serious problem. Columns and walls in compression, cantilever beams, and prestressed concrete structures that have not been designed according to specific seismic provisions are particularly susceptible to damage by vertical accelerations because they have little factor of safety against upward vertical acceleration. *Transfer girders*, horizontal members that support exterior perimeter columns in *tube buildings* (see Sec. 80) are definitely sensitive to vertical acceleration. The UBC covers these special cases in Sec. 2334(j) and, for dynamic analysis, in Secs. 2335(b)5 and 2335(e)4. (See Sec. 100.)

26 PROBABILITY OF OCCURRENCE[20]

The probability that an earthquake of magnitude M or greater will occur in a specific region in any given year is given approximately by Eq. 13.[21] B is a seismic

---

[20]Wiegel (1970) contains a more complete presentation of this subject.

[21]The form of Eq. 13 is easily derived from a Poisson distribution, which is commonly used to calculate the probability of an infrequent event.

parameter that has been approximately determined as 2.1 for the entire state of California and 0.48 for 100,000 mi^2 of southern California. While Eq. 13 does not place any upper bound on M, the probability of exceeding 8.5 is effectively zero.

$$p\{M\} = e^{-M/B} \qquad [13]$$

The expected number of earthquakes having magnitude greater than M during Y years is given by Eq. 14. For 100,000 mi^2 of southern California, C has a value of $1.7\,mi^{-2}$. Equation 14 is known as a *recurrence formula*. The quantity N/Y is the expected number of earthquakes per year. For northern California, $C = 76.7$ and $B = 0.847$, approximately. For the San Francisco area, $C = 19{,}700$ and $B = 0.463$, approximately.

$$N = CYe^{-M/B} \qquad [14]$$

27 FREQUENCY OF OCCURRENCE

For a specific area, an equation for the expected number, N, of earthquakes of a given magnitude, M, will be of the form

$$\log_{10} N = a - bM \qquad [15]$$

Taking the entire world as a whole, the approximate relationship (up to approximately $M = 8.2$) is

$$\log_{10} N = 7.7 - 0.9M \qquad [16]$$

Table 4 gives the expected number of earthquakes of any given magnitude per 100 years in California. (The table does not give the frequency over any particular location in the state.) Table 4 cannot be derived exactly from Eq. 16 because adjustments have been made to account for California's increased seismicity.

Table 4

Approximate Expected Frequency of Occurrence
of Earthquakes (per 100 Years)

Magnitude	Number
4.75–5.25	250
5.25–5.75	140
5.75–6.25	78
6.25–6.75	40
6.75–7.25	19
7.25–7.75	7.6
7.75–8.25	2.1
8.25–8.75	0.6

Data from G. W. Housner, "Strong Ground Motion," in *Earthquake Engineering*, Wiegel, ed., © 1970, p. 81. Reprinted by permission of Prentice-Hall, Inc., Englewood Cliffs, NJ.

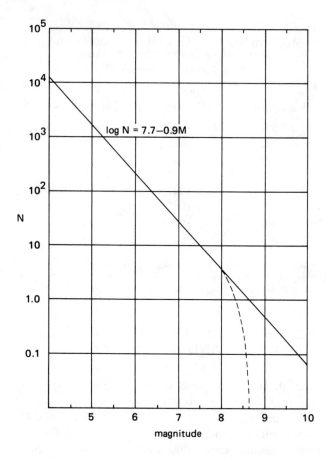

Figure 10   Expected Number of Earthquakes per Year

28 ATTENUATION OF GROUND MOTION

Ground motion at a site is related to the seismic energy received at that site, and when the propagation path is through rock, the amount of energy decreases the farther a site is from the epicenter.[22] This decrease is known as *attenuation*. Some of the factors affecting attenuation include path line, path length, focal depth, geological formations, properties of the crustal rock, and orientation of the fault.

Unfortunately, the geology and local conditions affect the actual values so much that little more than

---

[22]While the overall wave energy attenuates when the transmission path is through rock, the damage at a site does not necessarily decrease with distance. There are other factors that can concentrate the energy that reaches the site, as was proved by the Mexico City (1985) and Loma Prieta (1989) earthquakes. This is analogous to the decrease in the intensity of sunlight with distance from the sun. While the light may be diffused when it reaches the Earth's surface, it can be sufficiently concentrated with a lens to kindle a fire.

generalizations such as the following are possible about the rate of attenuation.[23]

◇ Intensity generally decreases with distance from the epicenter.

◇ There is little attenuation in the vicinity of the epicenter.

◇ Higher frequency components of the seismic wave attenuate faster than slower components do.

29 SEISMOMETER

Seismic waves (see Sec. 14) travel through the Earth and are recorded on seismometers. A *seismometer* is the detecting and recording parts of a larger apparatus known as a *seismograph*. Seismometers are pendulum-type devices and are mounted on the ground and measure the displacement of the ground with respect to a stationary reference point. Since a seismometer usually records motion in only one orthogonal direction, three seismometers are needed to record all components of ground motion. Figure 8 illustrates a typical seismometer trace, known as a *seismogram*. Appendix F is the actual seismogram of the 1940 El Centro earthquake.

Notice that while seismic activity usually continues for some time after the start of the earthquake, the major movement occurs in a concentrated period known as the *strong phase*. The longer the earthquake shakes, the more seismic energy is absorbed by buildings; thus, the duration of strong-phase shaking greatly affects the damage inflicted.[24] This is the reason that seismic engineers base their designs on the *effective peak ground acceleration* (see Sec. 38) rather than the actual peak ground acceleration: The peak acceleration is a spike that imparts no energy. The effective peak ground acceleration causes structures to move.

Seismometers record the varying amplitude of ground oscillations beneath the instrument. Sensitive seismometers greatly magnify these ground motions and can detect strong earthquakes occurring anywhere in the world. The time, location, and magnitude of an earthquake can be determined from the data recorded by seismometer stations.

Since a seismometer is a spring-mass-dashpot device, it will magnify or distort earthquakes with frequencies in certain ranges. The ratio of actual damping to critical damping can be changed to minimize such distortion. Good seismometer design calls for a damping ratio of between 0.6 and 0.7 with a natural period of vibration smaller than the smallest period to be measured.[25] (See Sec. 51 for damping ratio.)

30 ACCELEROMETER

An *accelerometer (accelerograph)* is a seismometer mounted in buildings for the purpose of recording large accelerations.[26] For this reason, they are also known as *strong motion seismometers*. The large swings accelerometers record typically exceed the scale limits of most seismometers. An accelerometer located in a building does not run continually. It is triggered by a P-wave (see Sec. 14) and runs for a fixed period of time.

Buildings located in seismic zones 3 and 4 over ten stories in height or over six stories and total floor areas of 60,000 ft^2 or more are required by the UBC [Sec. 2360, App. 23, Division II] to have at least three approved recording accelerographs.

31 OTHER SEISMIC INSTRUMENTS

A *tiltmeter* installed in the ground works on the same principle as a carpenter's level. The slightest movement of a bubble floating in a spherical dome is electronically detected to reveal tilting of the ground.

The strain (deformation) of rock under pressure can be measured by a *magnetometer*. Such strain changes the magnetic permeability of the rock, resulting in a local change in the magnetic field of the Earth.

Strain gauges measure how much the earth deforms. *Dilatometers* measure the earth's dilations. A dilatometer is a closed, fluid-filled tube approximately 3 m (10 ft)

---

[23]Numerous attenuation relationships have been published, particularly for sites and faults in California. However, there is little similarity between the correlations. Nevertheless, for sites located on firm soil, the attenuation laws are quite useful for predicting expected ground motions from future earthquakes.

[24]For example, the 1985 Chile earthquake (magnitude 7.8) had almost 80 s of strong ground motion. There were approximately 60 s of strong ground motion in the 1985 Mexico earthquake (magnitude 8.1). Both of these earthquakes resulted in significant loss of life and destruction. By comparison, the strong-phase motion of the 1940 El Centro earthquake (magnitude 7.1) lasted only 10 s, and the 1989 Loma Prieta earthquake (magnitude 7.1) had a mere 5 s. There is great debate and no consensus on whether or not long-duration earthquakes can occur in California.

[25]These design principles are incorporated into the Wood-Anderson seismometer, which has a damping ratio of approximately 0.8 (almost critical) and a natural period of 0.8 s (i.e., a natural frequency of 1.25 Hz).

[26]The distinction between the terms *seismometer* and *accelerometer* is not always made. However, it is important to recognize that seismometers typically run continually and record displacement, while accelerometers are triggered by P-waves and record acceleration.

long that is buried in the ground. Changes in the earth's "squeeze" are detected and measured by a pressure sensor or gauge at the top of the tube.

Scintillation counters are installed in wells to measure the amount of radioactive radon gas in the water. Minute amounts of radon are released into well water by rocks under stress.

Changes in the resistance of rock can be measured by a *resistivity gauge* and are indications of density and water content changes. Both density and water content change during periods of fluctuating stress.

A *creepmeter* measures minute gradual movement along a fault. In the past, such a meter relied on a wire stretched across a fault. Movement of the fault increased the tension in the wire. Current creepmeters use laser technology.

A *gravimeter* responds to variations in the local force of gravity. Such variations are the result of changes in underground rock density.

A *laser* can measure the round-trip travel time of a light beam between two points. When the relative positioning of the two points changes as a direct result of an earthquake, the travel time also changes.

32 EARTHQUAKE PREDICTION

While reliable long-term earthquake prediction remains elusive, short-term predictions based on the observation or measurement of various precursors (*premonitory signs*) seem possible. Most of the measuring devices mentioned in Secs. 29–31 can be adapted to instantaneous reporting. It may be possible to correlate sudden and unexpected changes in behavior (i.e., creep rate, tilt, accumulating strain, elevation, fluid pressure, seismic wave speed, electrical conductivity, and magnetic susceptibility) with the probability of an impending earthquake. Thus far, however, no reliable indicators have been found.

One proposed precursor, as related by the *seismic bay hypothesis*, may be the large-scale volume of the rock itself. As rock masses along a fault develop stresses, they crack and increase in volume. The volume increase is not consistently detectable, but in some cases, the volume increase can be observed.[27]

Another precursor is a decrease in the ratio of velocities of the two major seismic wave types. Normally, the ratio of velocities of the P- and S-waves is approximately

1.7, but this ratio decreases as the fault rock breaks. After a time, groundwater seeps in and fills the fissures (forming a "seismic bay"), and the velocity ratio returns to 1.7. According to the hypothesis, an earthquake can then occur.

Another possible precursor may be the emission of ultra-low-frequency radio waves before an earthquake.[28] Additional research in this area is required.

33 OTHER EARTHQUAKE CHARACTERISTICS

In addition to the peak ground acceleration, two other characteristics that contribute significantly to the effects of an earthquake are its duration of strong shaking (motion) and frequency content.[29] Roughly speaking, the longer the duration of strong shaking, the greater the energy that can be imparted to a structure. Since various parts of a structure can absorb only a limited amount of elastic strain energy, a longer earthquake has a greater chance of driving structural performance into inelastic behavior.

The shaking in the 1989 Loma Prieta earthquake lasted approximately 10–15 s. It is generally believed that a longer duration (i.e., another 20 s) would have resulted in significantly greater damage. The UBC provisions are intended to accommodate earthquakes with durations of 10–60 s.

The effects of resonance on all types of machinery and structures are well known. Basically, if a regular disturbing force is applied at the same frequency as the natural frequency (see Sec. 49), the oscillation of the structure can be greatly magnified. In such cases, the effects of damping are minimal. While earthquakes are never as regular as a sinusoidal waveform, there is usually a predominant waveform that is roughly regular.[30]

---

[27]The *Palmdale bulge* is thought to be an example of this volume increase.

[28]Ultra-low-frequency (ULF) radio waves (below 5 Hz) were received from the vicinity of the epicenter in the October 1989 Loma Prieta earthquake. However, a direct cause-and-effect relationship has yet to be proved.

[29]Note that this characteristic is defined as the duration of strong shaking, not the overall duration of the earthquake.

[30]Fourier analysis can be used to separate the dominant frequencies (i.e., the frequencies at which most seismic energy arrives at a site) of a specific earthquake. However, this is rarely done for most seismic design projects.

3

EFFECTS OF EARTHQUAKES ON STRUCTURES

34 SEISMIC DAMAGE

Structural damage due to an earthquake is not solely a function of the earthquake ground motion. The primary factors affecting the extent of damage are

◇ *earthquake characteristics*, such as (a) peak ground acceleration, (b) duration of strong shaking, (c) frequency content, and (d) length of fault rupture

◇ *site characteristics*, such as (a) distance between the epicenter and structure, (b) geology between the epicenter and structure, (c) soil conditions at the site, and (d) natural period of the site

◇ *structural characteristics*, such as (a) natural period and damping of the structure, (b) age and construction method of the structure, and (c) seismic provisions (i.e., detailing) included in the design

35 SEISMIC RISK ZONES

In order to design a structure to withstand the effects of an earthquake, it is necessary to determine the expected earthquake magnitude. While extensive mathematical models could be developed for each location, seismic codes have evolved a simplified model based on *seismic zones*. This model is based on the seismic probability map shown in Fig. 11.[31]

There are several methods of evaluating the significance of the seismic risk zones. One method is to correlate the zones with the approximate accelerations and magnitudes, as Table 5 does.

---

[31]The map shown in Fig. 11 is subject to change with different versions of the seismic code.

Table 5

Approximate Code Maximum Zone Acceleration and Magnitude

Zone	Maximum Acceleration	Maximum Magnitude
0	0.04 g	4.25
1	0.08 g	4.75
2	0.16 g	5.75
3 (not near a great fault)	0.33 g	7.00
4 (near a great fault)	0.50 g	8.5

Another interpretation of the significance of the zones is to correlate them to the effects of an earthquake and the Modified Mercalli intensity scale.

Table 6

Effects of an Earthquake by Zone

Zone	Effect
0	no damage
1	minor damage corresponding to MM intensities V and VI; distant earthquakes may damage structures with fundamental periods greater than 1.0 s
2	moderate damage corresponding to MM intensity VII
3	major damage corresponding to MM intensity VIII and higher
4	same as zone 3

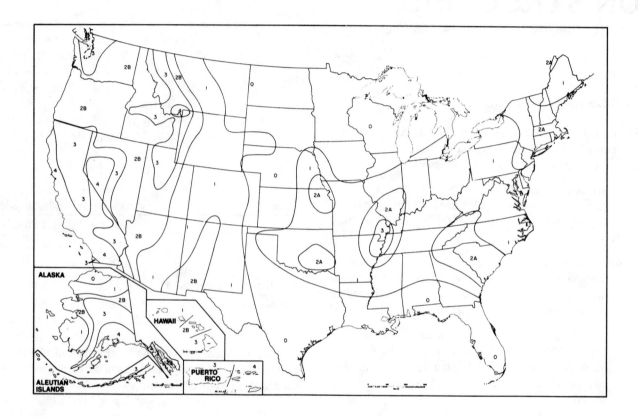

Figure 11   Seismic Risk Zones

Reproduced from the 1991 edition of the *Uniform Building Code*, copyright © 1991,
with the permission of the publishers, the International Conference of Building Officials.
[UBC Fig. 23-2]

36 RISK MICROZONES

Certain limited areas, referred to as *microzones*, consistently experience higher ground accelerations than do surrounding areas.[32] This tendency is primarily attributed to the site (that is, soil) conditions in the microzone, as the soil profile affects the peak ground acceleration, frequency content, and duration of strong motion. Inasmuch as seismic damage is at least partially related to ground acceleration, knowledge of such microzonification is essential. The *microzonification* concept, however, has not been explicitly incorporated in the Uniform Building Code, although it is implicit in the design process.

Mexico City has incorporated microzones into its rebuilding plan following the devastating earthquake in 1985. Considering the significant variations in damage in different areas of the San Francisco Bay Area during the Loma Prieta earthquake, microzonification has also been widely proposed. (For example, unreinforced buildings in Chinatown were not damaged, while similarly-constructed buildings in the Marina district were. Portions of the Cypress structure on clay collapsed, but other portions built on firmer soil did not.)

37 PROBABLE AND CREDIBLE EARTHQUAKE GROUND MOTIONS

A *maximum probable earthquake* ground motion at a site is the largest earthquake shaking that has a significant probability of occurring within the lifetime of a structure due to earthquakes from all sources. Since the potential losses of property and life are very great, the probability does not have to be very large to be significant.

The *maximum credible earthquake* ground motion at a site is the maximum possible earthquake ground motion based on available knowledge of the building site location. The maximum credible earthquake is difficult to evaluate. However, energy requirements seem to limit the maximum worldwide Richter magnitude to 9.0 or less. In California, the maximum is generally considered to be 8.5.

38 EFFECTIVE PEAK GROUND ACCELERATION

The *effective peak ground acceleration*, EPA (or A_a as it is used in some documents), is the maximum acceleration to which a building responds (i.e., produces an increase in response) and is more of a contrived design parameter set by code than a feature of an actual earthquake.[33] As a code provision, EPA depends on the region and corresponds numerically in gravities to the Z coefficient (see Sec. 91) in the UBC.

As explained in the Blue Book commentary, Sec. 1D.2, the EPA is derived from the log-tripartite graph given in ATC 3-06 scaled downward from the spectral acceleration by dividing it by a spectral amplification factor. (See Sec. 42 for spectral acceleration.) The value of the spectral amplification factor depends on the amount of damping present in the building and the probability of the earthquake's occurrence. For example, for 5% damping and a hazard level (probability of occurrence) of 10% in a 50-year period, the spectral amplification factor is 2.5.

The method of deriving the EPA is subject to continuing study and analysis. In general, the EPA is usually somewhat less than the PGA.[34]

39 SITE PERIOD

The *site (soil) period* is now recognized as a significant factor contributing to structural damage.[35] When a site has a natural frequency of vibration that corresponds to the predominant earthquake frequency, site movement can be greatly magnified. This is known as *resonance*. (See Sec. 57.) Thus, the buildings can experience ground motion much greater than would be predicted from only the seismic energy release.

Determining the actual site period is no easy matter. Since the site period can be computed precisely from

---

[32] For example, in the 1989 Loma Prieta earthquake, peak ground accelerations in San Francisco did not generally exceed 0.09 g. However, a 31-story instrumented office building in Emeryville experienced a horizontal acceleration of 0.26 g at the ground. Peak accelerations at the Bay Bridge are believed to have ranged between 0.22 g and 0.33 g. The Golden Gate Bridge experienced 0.24 g, while 0.33 g was recorded at the San Francisco Airport. Similar ground accelerations were recorded near the collapsed Interstate 880 Cypress structure.

[33] The terms *effective peak ground acceleration* (EPA) and *effective peak velocity* (EPV) were originally defined in the commentary of ATC 3-06. See Sec. 82.

[34] By designing buildings for lower than the PGA, the designer counts on the ductility of the structure to dissipate the seismic energy in excess of the energy required to stress the structure to its elastic point. Although the designer does not specifically design to make the structure yield, the net result is to ensure yielding (a seismic energy dissipation mechanism). Yielding, however, is not synonymous with collapse.

[35] The resonance-induced damages of the 1985 Mexico City earthquake and the 1989 Loma Prieta earthquake are prime examples.

widely available formulas and still be grossly inaccurate, such determinations are best left to experts familiar with the area.

Other soil characteristics, including density, bearing strength, moisture content, compressibility (i.e., tendency to settle), and sensitivity (i.e., tendency to liquefy), are additional factors not addressed by the seismic code. These factors, nevertheless, must be considered in structural design.

40 SOIL LIQUEFACTION

Liquefaction occurs in soils, particularly in soils of saturated cohesionless particles such as sand, and is a sudden drop in shear strength. This is experienced as a drop in bearing capacity. In effect, the soil turns into a liquid, allowing everything it previously supported to sink. It is not necessary for the soil to be located on a cliff or other escarpment for liquefaction to occur. Perfectly flat soil layers can become major mud puddles if the conditions are right.[36]

Continued cycles of reversed shear in saturated sand can cause *pore water pressure* to increase, which in turn decreases the *effective stress* and *shear strength*.[37] When the shear strength drops to zero, the sand liquefies.

Conditions most likely to contribute to or indicate a potential for liquefaction include (1) a lightly-loaded sand layer within 15 to 20 m of the surface, (2) uniform particles of medium size, (3) a saturated condition below a water table, and (4) a low penetration-test value.

41 BUILDING PERIOD

When a lightly-damped building is displaced laterally by an earthquake, wind, or other force, it will oscillate back and forth with a regular *period*. (This building period should not be confused with the site period mentioned in Sec. 39 or with the period of the earthquake.)

The natural period of modern buildings can seldom, if ever, be calculated from simple vibrational theory. Four other methods, listed below, can be used when knowledge of a building period is needed.

1. Analytical models based on finite element analysis (FEA) and other modeling techniques can be used.

2. A scale model of the building can be constructed and the natural period extrapolated from measurements on the model. (This is seldom done, however.)

3. If the building has been constructed, actual measurements can be taken.

4. Empirical relations (such as are incorporated in the UBC Sec. 2334(b)) can be used. (See Sec. 94.)

42 SPECTRAL CHARACTERISTICS

Despite some inherent regularity, earthquake seismograms are quite "noisy." It is difficult to determine how a building behaves at all times during an earthquake consisting of many random pulses. It is also unnecessary in many cases to know the entire time-history response of the building, since the maximum seismic force on (and, hence, damage in) a structure depends partially on the effective peak acceleration experienced, not on lower accelerations that might have occurred during the earthquake.

The maximum acceleration[38] that is experienced by a single-degree-of-freedom vibratory system (see Sec. 46) is known as the *spectral acceleration*, S_a.[39] Similarly, the maximum displacement and velocity are known as the *spectral displacement*, S_d, and *spectral velocity*, S_v, respectively.

---

[36]Dramatic examples of liquefaction occurred in the 1964 earthquakes in Alaska and Niigata, Japan.

[37]These are standard terms used in soils and foundations handbooks.

[38]The maximum building acceleration should not be confused with the effective ground acceleration. The building acceleration is typically higher than the ground acceleration. The ratio of building to ground acceleration depends on the building period, a concept that is discussed elsewhere in this book. For infinitely stiff buildings (with zero natural periods), the ratio is 1. The spectral acceleration from typical California design earthquakes (i.e., those used as a basis in establishing the UBC provisions) for a 10% damped building located on rocks or other firm soil is approximately 2.0 to 2.5 times the peak ground acceleration. (See Sec. 38.)

[39]Another name is *spectral pseudo acceleration*—"pseudo" because the value does not correspond exactly to the maximum acceleration.

43 BASE SHEAR

The theoretical maximum seismic force, V, on a structure of mass m (weight W) is known as the *base shear*, and is given by Newton's second law ($F = ma$).[40] Note that while the spectral acceleration is sometimes given in fractions or percentages of gravity (g) or normalized in some other manner, Eq. 17 requires the spectral acceleration to be expressed in ft/sec^2, in/sec^2, or m/s^2.

$$V = mS_a = \frac{WS_a}{g} \quad \text{[SI]} \qquad \text{[17(a)]}$$

$$V = \frac{mS_a}{g_c} = \frac{WS_a}{g} \quad \text{[U.S.]} \qquad \text{[17(b)]}$$

44 RELATIONSHIP BETWEEN SPECTRAL VALUES

Equation 18 indicates that the spectral displacement, velocity, and acceleration can be derived from one another if the natural frequency (in rad/s) of vibration, ω, is known.[41] Equation 18 is exact for the case of an undamped, single-degree-of-freedom system in simple harmonic motion but is approximate otherwise (i.e., is approximate with damping and for multiple-degree-of-freedom systems). (See Sec. 46.)

$$|S_d| = \left|\frac{S_v}{\omega}\right| = \left|\frac{S_a}{\omega^2}\right| \quad \text{[undamped SDOF]} \qquad \text{[18]}$$

---

[40]The total building dead load (dead weight) is used in the calculation of the base shear. This practice should not be confused with calculations of diaphragm force which omit half of the ground floor dead load. Except for a warehouse, no live load is included. While these may seem like arbitrary provisions, the UBC is specific in including and excluding certain fractions of the building weight and live load. (See Sec. 97.)

[41]Equation 18 is easily derived for the case of sinusoidal oscillation. Starting with a sinusoidal position equation, $x(t) = A \sin \omega t$, the first derivative (i.e., the velocity equation) is $v(t) = \omega A \cos \omega t$, whose maximum amplitude is ω multiplied by the amplitude of the position function. The maximum acceleration amplitude is similarly determined.

4
VIBRATION THEORY

45 TWO APPROACHES TO SEISMIC DESIGN

There are two greatly different approaches to seismic design, both of which are "correct" in their own ways. In a *dynamic analysis*, the overall building and story stiffnesses and rigidities are calculated. (See Sec. 48.) A specific design earthquake, including magnitude and loading history, is selected and applied to a mathematical model (consisting of lumped masses, damping, and spring stiffness) of the building. The solution may rest heavily on vibrational theory, finite element analysis, and other advanced structural techniques requiring computer analysis. The response of the system (including the displacement and acceleration functions) is calculated and used to determine the forces in each member as a function of time. This method is now almost always used for critical structures such as dams and power plants.

There are a number of factors that can render the dynamic approach inappropriate. The building itself may be too simple or too standardized to warrant the rigorous approach of the design analysis. Conversely, the building may be too complex and have too many degrees of freedom to model mathematically. Also, in the initial design phases, the member sizes and locations may not be known, making it difficult to estimate stiffnesses and rigidities.[42] The dynamic approach is inappropriate, too, when the design earthquake is not known. Additionally, the analysis may be beyond the financial or computational abilities of the engineering firm performing the design. And, finally, unless the building is particularly irregular as defined in UBC Sec. 2333(e)3, there may be no code requirement to perform a dynamic analysis.

The alternative to a dynamic analysis is a *static analysis*. The *equivalent lateral (seismic) force* is calculated as simply some fraction of the dead weight. Chapter 23 of the UBC codifies this analysis so that there is no need to know the design earthquake.

In the chapters and sections that follow, these two methods are at times discussed separately and, at other times, aspects of each method are combined. Although considerably different in approach, the static method is based on engineering logic that can, in many cases, be traced back to vibration theory.

46 SIMPLE HARMONIC MOTION[43]

Ideal vibrational systems that consist of springs and masses and that are not acted upon by external disturbing forces (after an initial displacement) are known as *simple harmonic oscillators*. During steady-state motion, such oscillators move in a repetitive sinusoidal pattern known as *simple harmonic motion*. Simple harmonic motion is characterized by the absence of a continued disturbing force and a lack of frictional damping.

Examples of simple harmonic oscillators are a mass hanging on an ideal spring (Fig. 12(a)), a pendulum on a frictionless pivot, and a slab supported on two massless cantilever springs (Fig. 12(b)).

---

[42]Although there are some real design programs, most "design" programs are actually analysis programs that require the user to input information about the locations and characteristics of the structural members.

[43]There is no need actually to develop the differential equations of oscillatory motion for a building. However, this section introduces some of the terms and concepts related to structural dynamics.

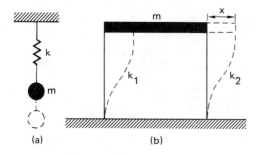

Figure 12   Simple Harmonic Oscillator

The number of variables needed to define the position of all parts of a system is known as the *degree of freedom*. If the oscillator is constrained to move in one dimension only, or alternatively, if one linear or angular variable is sufficient to describe the position of the oscillator, the system is known as a *single-degree-of-freedom* (SDOF) *system*. The moving mass in an SDOF system is usually concentrated at one point and is known as a *lumped mass*.

Oscillation of the SDOF system shown in Fig. 13 is initiated by displacing and releasing the mass. The displacement, x, is measured from the equilibrium position. Once the system has been displaced and released, no further external force acts on it. Because there is no friction once it is set in motion, the mass remains in motion indefinitely.

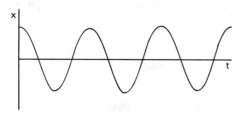

Figure 13   Time Response of Simple Harmonic Oscillator

47 STIFFNESS AND FLEXIBILITY

When a force, F, acts on an ideal linear spring, *Hooke's law* predicts the magnitude of the spring deflection, x. In Eq. 19, k is the *stiffness* or *spring constant* in N/m (lbf/ft). The stiffness is the force that must be applied in order to deflect the spring a distance of one unit.

$$F = kx \quad \text{[Hooke's law]} \qquad [19]$$

Referring to the mass-spring system shown in Fig. 12(a), the spring is undeflected until the mass is attached to

it. After the mass is attached, the spring will deflect an amount known as the *static deflection*, x_{static}.

$$W = mg = kx_{\text{static}} \qquad \text{[SI]} \quad [20(a)]$$

$$W = \frac{mg}{g_c} = kx_{\text{static}} \qquad \text{[U.S.]} \quad [20(b)]$$

The stiffness, k, of a beam can be calculated as the ratio of applied force to deflection from the beam deflection tables that are typically in every mechanics of materials textbook. Table 7 summarizes some of these terms.

Flexibility is the reciprocal of stiffness. It is the deflection obtained when a unit force is applied. Therefore, its units are m/N (ft/lbf).

48 RIGIDITY

Strictly speaking, *rigidity*, R, is the reciprocal of deflection. In buildings where all members consist of the same material and all walls have the same thickness (for example, a masonry-walled building or an all-concrete building), the deflection is traditionally calculated with arbitrary values of applied force, modulus of elasticity, and wall thickness. This is permitted when distributing the applied lateral loads to vertical members because the load "taken" by each member is proportional to the member's *relative rigidity*. (See also Sec. 119.)

$$R = \frac{1}{x} \qquad [21]$$

Both moment and shear contribute to the deflection experienced by a vertical member (e.g., a shear wall).[44] Consider the wall shown in Fig. 14(a). This wall is fixed at the top and bottom and bends in double curvature since the top and bottom must remain vertical. Such a wall is known as a *fixed pier*. The deflection due to both shear, V, and moment, Vh, of a fixed pier is given by Eq. 22.

$$x_{\text{fixed}} = \frac{Fh^3}{12EI} + \frac{1.2Fh}{AG} \quad \text{[fixed pier]} \qquad [22]$$

$$A = td \qquad [23]$$

$$I = \frac{td^3}{12} \qquad [24]$$

---

[44]Common beam deflection equations, such as those presented in Table 7, usually disregard the effect of shear. However, shear contributes to deflection, particularly in materials such as concrete and masonry that are not strong in shear (i.e., have low shear moduli) and when the ratio of height to depth is low. In general, shear deflection should not be neglected.

Table 7
Deflection and Stiffness for Various Systems
(Due to Bending Moment Alone)

System	Maximum Deflection (x)	Stiffness (k)
	$\dfrac{Fh}{AE}$	$\dfrac{AE}{h}$
	$\dfrac{Fh^3}{3EI}$	$\dfrac{3EI}{h^3}$
	$\dfrac{Fh^3}{12EI}$	$\dfrac{12EI}{h^3}$
	$\dfrac{wL^4}{8EI}$	$\dfrac{8EI}{L^3}$
	$\dfrac{Fh^3}{12E(I_1+I_2)}$	$\dfrac{12E(I_1+I_2)}{h^3}$
	$\dfrac{FL^3}{48EI}$	$\dfrac{48EI}{L^3}$
(w is load per unit length)	$\dfrac{5wL^4}{384EI}$	$\dfrac{384EI}{5L^3}$
	$\dfrac{FL^3}{192EI}$	$\dfrac{192EI}{L^3}$
(w is load per unit length)	$\dfrac{wL^4}{384EI}$	$\dfrac{384EI}{L^3}$

A wall that is fixed at the bottom but free to rotate at the top bends in simple curvature and is known as a *cantilever pier*. The deflection of a cantilever wall due to both effects is

$$x_{\text{cantilever}} = \frac{Fh^3}{3EI} + \frac{1.2Fh}{AG} \quad \text{[cantilever pier]} \quad [25]$$

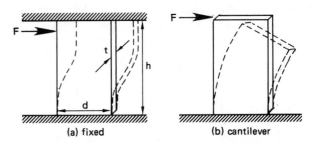

(a) fixed   (b) cantilever

Figure 14   Fixed and Cantilever Piers

For concrete, $E \approx 3 \times 10^6$ psi $(2.1 \times 10^7 \text{ kPa})$ and $G \approx 0.4E$. For masonry, $E \approx 1 \times 10^6$ psi $(6.9 \times 10^6 \text{ kPa})$ and $G \approx 0.4E$. For steel, $E \approx 3 \times 10^7$ psi $(2.1 \times 10^8 \text{ kPa})$ and $G \approx 1.2 \times 10^6$ psi $(8.3 \times 10^7 \text{ kPa})$. However, since the shear that is distributed to each vertical member (i.e., each pier) is in proportion to the relative rigidity and does not depend on the actual rigidity, the deflections can be calculated with arbitrary values of total shear, F, and wall thickness, t. Equations 26 and 27 use $F = 100,000$, $t = 1.0$, $E = 1,000,000$, and arbitrary units.

$$R_{\text{fixed}} = \frac{1}{0.1\left(\dfrac{h}{d}\right)^3 + 0.3\left(\dfrac{h}{d}\right)} \quad [26]$$

$$R_{\text{cantilever}} = \frac{1}{0.4\left(\dfrac{h}{d}\right)^3 + 0.3\left(\dfrac{h}{d}\right)} \quad [27]$$

49 NATURAL PERIOD AND FREQUENCY

The time for a complete cycle of oscillation of an SDOF system is known as the *natural period*, T, usually expressed in seconds. The reciprocal of natural period is the *linear natural frequency*, f, usually called *natural frequency* or just *frequency*, and is expressed in Hz (i.e., cycles per second). It is important to distinguish between the natural frequency of a system (building, oscillator, etc.) and the frequency of an applied force. The natural frequency, f, in Eq. 28 has nothing to do with an external force.

$$f = \frac{1}{T} \quad [28]$$

The natural frequency can also be expressed in radians per second (rad/s), in which case it is known as the *circular frequency, angular natural frequency,* or just *angular frequency, ω*.

$$\omega = 2\pi f = \frac{2\pi}{T} \qquad [29]$$

It is easy to derive the natural frequency for the case of a simple harmonic oscillator.[45] For a mass on a spring,

$$\omega = \sqrt{\frac{k}{m}} \qquad \text{[SI]} \quad [30(a)]$$

$$\omega = \sqrt{\frac{kg_c}{m}} = \sqrt{\frac{kg}{W}} \qquad \text{[U.S.]} \quad [30(b)]$$

Substituting k from Hooke's law (Eq. 19) and recognizing that the mass, m, can be calculated from the weight, W, an expression is derived for the natural frequency in terms of the static deflection, x_{st}, calculated in Sec. 47.

$$\omega = \frac{2\pi}{T} = \sqrt{\frac{F}{x_{st}m}} \qquad \text{[SI]} \quad [31(a)]$$

$$\omega = \frac{2\pi}{T} = \sqrt{\frac{Fg_c}{x_{st}m}} = \sqrt{\frac{Fg}{x_{st}W}} \qquad \text{[U.S.]} \quad [31(b)]$$

Since Eq. 31 can be used to calculate the natural period, it is tempting to substitute the maximum allowable code drift (i.e., 0.5% of the total building height; see Sec. 107) for the static deflection in order to calculate the maximum natural building period.[46] Such a substitution would require no structural analysis at all but implies that the building will have maximum flexibility permitted by the code. The problem with this approach is that it assumes the maximum allowable drift to be the same for all zones, although the flexibility actually depends on the zone since flexibility is affected by the building's seismic resistance. Thus, while the lateral forces on the building differ, the maximum drift and thus, the period do not. Obviously, something is wrong with calculating the building period in this way.

---

[45]This is done in virtually every physics, dynamics, and earthquake book, but not here.

[46]In fact, the commentary for the 1985 Blue Book discussed using two-thirds of the maximum allowable drift to do just this very thing.

Example 1

A 2.23-kg (5-lbm) mass hangs from two ideal springs as shown. (Assume the block is constrained so that it does not rotate.) What is the natural period of vibration?

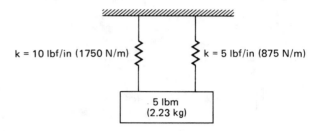

SI Solution

Both springs must deflect in order for the mass to move. The total composite spring constant is

$$k_t = k_1 + k_2 = 1750\,\frac{\text{N}}{\text{m}} + 875\,\frac{\text{N}}{\text{m}}$$

$$= 2625\,\text{N/m}$$

From Eqs. 28 and 30, the natural period for vertical translation is

$$T = 2\pi\sqrt{\frac{m}{k}} = 2\pi\sqrt{\frac{2.23\,\text{kg}}{2625\,\frac{\text{N}}{\text{m}}}}$$

$$= 0.183\,\text{s}$$

Customary U.S. Solution

Both springs must deflect in order for the mass to move. The total composite spring constant is

$$k_t = k_1 + k_2 = \left(5\,\frac{\text{lbf}}{\text{in}} + 10\,\frac{\text{lbf}}{\text{in}}\right)\left(12\,\frac{\text{in}}{\text{ft}}\right)$$

$$= 180\,\text{lbf/ft}$$

From Eqs. 28 and 30, the natural period is

$$T = 2\pi\sqrt{\frac{m}{g_c k}}$$

$$= 2\pi\sqrt{\frac{5\,\text{lbm}}{\left(32.2\,\frac{\text{ft-lbm}}{\text{lbf-sec}^2}\right)\left(180\,\frac{\text{lbf}}{\text{ft}}\right)}}$$

$$= 0.185\,\text{sec}$$

Example 2

A small water tank is supported on a slender column as shown. Neglecting the weight of the column, calculate the natural period of vibration.

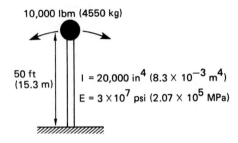

10,000 lbm (4550 kg)

50 ft (15.3 m)

$I = 20,000 \text{ in}^4 \, (8.3 \times 10^{-3} \text{ m}^4)$

$E = 3 \times 10^7 \text{ psi} \, (2.07 \times 10^5 \text{ MPa})$

SI Solution

Consider the water tower to be a cantilever beam. The stiffness is the lateral (i.e., sideways) deflection when a unit force is applied.

$$k = \frac{3EI}{h^3}$$

$$= \frac{(3)(2.07 \times 10^5 \text{ MPa})\left(10^6 \, \dfrac{\text{Pa}}{\text{MPa}}\right)(8.3 \times 10^{-3} \text{ m}^4)}{(15.3 \text{ m})^3}$$

$$= 1.44 \times 10^6 \text{ N/m}$$

From Eq. 30, the period is

$$T = 2\pi\sqrt{\frac{m}{k}} = 2\pi\sqrt{\frac{4550 \text{ kg}}{1.44 \times 10^6 \, \dfrac{\text{N}}{\text{m}}}}$$

$$= 0.35 \text{ s}$$

Customary U.S. Solution

Consider the water tower to be a cantilever beam. The stiffness is the lateral (i.e., sideways) deflection when a unit force is applied.

$$k = \frac{3EI}{h^3} = \frac{(3)\left(3 \times 10^7 \, \dfrac{\text{lbf}}{\text{in}^2}\right)(20,000 \text{ in}^4)}{(50 \text{ ft})^3 \left(12 \, \dfrac{\text{in}}{\text{ft}}\right)^2}$$

$$= 1 \times 10^5 \text{ lbf/ft}$$

From Eq. 30, the period is

$$T = 2\pi\sqrt{\frac{m}{g_c k}}$$

$$= 2\pi\sqrt{\frac{10,000 \text{ lbm}}{\left(32.2 \, \dfrac{\text{ft-lbm}}{\text{lbf-sec}^2}\right)\left(1 \times 10^5 \, \dfrac{\text{lbf}}{\text{ft}}\right)}}$$

$$= 0.35 \text{ sec}$$

50 DAMPING

Damping is the dissipation of energy from an oscillating system, primarily through friction. The kinetic energy is transformed into heat. All structures have their own unique ways of dissipating kinetic energy, and in certain designs, mechanical systems known as *dampers* (see Sec. 155) can be installed to increase the damping rate.[47]

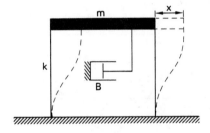

Figure 15   Oscillator with Damping

There are several sources of damping. *External viscous damping* is caused by the structure moving through surrounding air (or water, in some cases). It is generally small in comparison to other sources of damping. *Internal viscous damping*, commonly the only type of damping actually modeled, is related to the viscosity of the structural material. It is proportional to velocity. (See Eq. 32.) *Body-friction damping*, also known as *Coulomb friction*, results from friction between members in contact. It includes friction at connection points. Sections of opposed cracked masonry walls rubbing back and forth against one another are very effective body-friction dampers. Another source of damping, *radiation damping*, occurs as a structure vibrates and becomes a source of energy itself. Some of the energy is reradiated

---

[47]A damper is similar in design to a shock absorber and is often depicted as a plunger moving through a pot of viscous fluid. In modeling, dampers are also known as *dashpots*, although this term is more common among mechanical engineers.

through the foundation back into the ground. Finally, *hysteresis damping* occurs when the structure yields during reversals of the load. (See Sec. 71.)

For internal viscous damping, the frictional damping force opposing motion is given by Eq. 32. The exponent n in Eq. 32 is usually taken as 1.0 for slow-moving systems and 2.0 for fast-moving systems. However, even these values are idealizations. The coefficient B in Eq. 32 is known as the *damping coefficient*.

$$F_{\text{damping}} = Bv^n \qquad [32]$$

51 DAMPING RATIO

An oscillating system with a small amount of damping will continue to oscillate, although the amplitude of the oscillations will decay. Many cycles and a long time may elapse before the system eventually reaches the motionless equilibrium position. This type of system is known as an *underdamped system*.

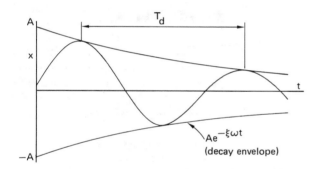

Figure 16   Underdamped Motion
(Moderate Damping)

Conversely, a system may have a large amount of damping. When displaced, such an *overdamped system* seems to "hang in space," taking an extremely long time to return to the motionless equilibrium position.[48]

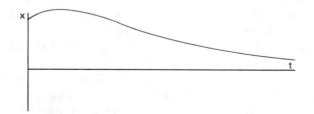

Figure 17   Overdamped Motion

---

[48]An example of an overdamped system is a door with a slow-closing device that will not permit the door to slam shut. Instead, the door approaches the fully-closed position slowly.

Both the underdamped and overdamped cases bring the system back to the equilibrium position only after a long time. There is one particular amount of damping, known as *critical damping*, that brings the system to equilibrium in a minimum time without oscillation. In this case, the damping coefficient, B, is known as the *critical damping coefficient*, B_{critical}.

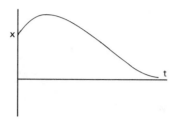

Figure 18   Critically-Damped Motion

Most systems are not critically damped. The ratio of the actual damping coefficient to the critical damping coefficient is known as the *damping ratio*, ξ.

$$\xi = \frac{B}{B_{\text{critical}}} \qquad [33]$$

52 DECAY ENVELOPE

For small and moderate amounts of damping (i.e., the underdamped case), the oscillation will be bounded by a *decay envelope* as was illustrated in Fig. 16. The equation of the decay envelope is given by Eq. 34.

$$x = Ae^{-\xi\omega t} \qquad [34]$$

The ratio of one cycle's amplitude to the subsequent cycle's amplitude is the *decay decrement*. The natural logarithm of the decay decrement is the *logarithmic decrement*, δ.

$$\delta = \ln\left(\frac{x_n}{x_{n+1}}\right) = \frac{2\pi\xi}{\sqrt{1-\xi^2}} \qquad [35]$$

53 DAMPING RATIO OF BUILDINGS

The exact damping ratio, ξ, of an actual structure is difficult to determine. Furthermore, the damping ratio increases during large swings. Available data on actual structures suggest the values given in Table 8. There is little evidence to support damping ratios in real structures that exceed 15%.

Table 8
Typical Damping Ratios

Type of Construction	ξ

steel frame
   welded connections
   flexible walls
	0.02

steel frame
   welded connections
   normal floors
   exterior cladding
	0.05

steel frame
   bolted connections
   normal floors
   exterior cladding
	0.10

concrete frame flexible internal walls	0.05

concrete frame
   flexible internal walls
   exterior cladding
	0.07

concrete frame
   concrete or masonry
     shear walls
	0.10

concrete or masonry shear wall	0.10
wood frame and shear wall	0.15

Although the damping ratio is essentially constant for a given building, the damping ratio of a particular building type or construction material appears to depend on the natural period of the building. Buildings with natural periods of less than 1.0 s may have damping ratios two to three times higher than buildings with similar construction but natural periods greater than 1.0 s. While generalizations that do not consider all factors are possible, it appears that the building's damping ratio, period, and construction method are all related.

54 DAMPED PERIOD OF VIBRATION

The period of oscillation of a system will be slightly greater with damping than without it, since the damping slows down the movement. Equations 36 and 37 give the damped frequency and period. Most buildings have only small amounts of damping. Therefore, the damped and undamped periods are almost identical.

$$\omega_d = \omega \sqrt{1 - \xi^2} \qquad [36]$$

$$T_d = \frac{2\pi}{\omega_d} \qquad [37]$$

55 FORCED SYSTEMS

A *forced system* is an oscillatory system that is supplied energy on a regular, irregular, or random basis. The force that supplies the energy is known as a *forcing function*. Forcing functions can be constant (i.e., a *step function*), applied and quickly removed (i.e., an *impulse function*), sinusoidal, or random.

An example of a regularly-forced system is a flexible floor supporting an out-of-balance motor. When turning, the motor will generate a force at a frequency proportional to the motor's rotational speed. An example of a randomly-forced system is a structure acted on by wind or seismic forces. In the latter case, there is little or no regularity to the applied forces.

It is not significant whether a lateral force (e.g., seismic force or wind) is applied to a building directly or whether the base moves out from under the building (e.g., as in an earthquake). In the latter case, the equivalent lateral force is an inertial force, but it is just as effective at displacing the building relative to its base as any direct force is.

The system response (i.e., the behavior of a building) to a force depends on the nature of the forcing function. Unfortunately, earthquakes are never simple sinusoids and buildings have more than a single degree of freedom (see Sec. 60), so the determination of system response is time-consuming and complex. Computers and numerical techniques, however, greatly simplify the analysis.[49]

56 MAGNIFICATION FACTOR

It is not difficult to show that when a sinusoidal forcing function with the form $F(t) = P \sin \omega_f t$ is applied to a system with stiffness k, the steady-state response will be of the form of Eq. 38.

$$x(t) = \beta \left(\frac{P}{k} \right) \sin \omega_f t \qquad [38]$$

In Eq. 38, P/k is the *static deflection*, x_{static}, (see Sec. 47) that is experienced if a constant force P is applied to the system. β is a dynamic *magnification factor* that depends on all other characteristics of the system.[50]

---

[49]It may not always be a simple matter, however, to interpret the results of the analysis.

[50]The dynamic magnification factor depends on the natural and forcing frequencies, the mass in motion, and the amount of damping (or, alternatively, on the damping ratio). Formulas for calculating the magnification factor for damped and undamped cases are given in virtually every textbook covering vibration theory.

57 RESONANCE

For a given system, the dynamic magnification factor, β, can be less than or greater than unity, depending on the ratio of the natural and forcing frequencies. Figure 19 illustrates how the magnification factor varies for different frequency ratios. At one point, corresponding to where the forcing function frequency equals the natural frequency of the system, the magnification factor is very large (theoretically infinite for undamped systems). Such a condition is known as *resonance*. The ratio ω_f/ω must be greater than $\sqrt{2}$ for β to drop below $|1.0|$.

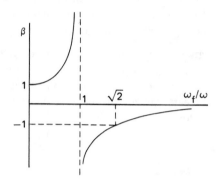

Figure 19   Undamped Magnification Factor

The 1985 Mexico City magnitude 8.1 earthquake occurred on September 19, with a 7.5 aftershock occurring the next day. Approximately 400 buildings were destroyed, and 700 were damaged. The death toll was over 5000. The earthquake consisted of (approximately) twenty 0.18 g pulses coming every 2 s (the natural period of the ground). This coincided with the period for buildings in the 7–20 story range. The resulting resonance-related yielding was the primary cause of structural failure. Quality of construction was not a major factor in the widespread destruction.

Resonance is now considered a prime factor in the collapse of the Oakland Interstate 880 Cypress structure during the October 17, 1989 Loma Prieta earthquake. The structure had a natural frequency of 2–4 Hz, which coincided with the 3–5 Hz natural period of the deep mud that underlaid piles that supported portions of the freeway that collapsed. The depth of the mud and the length of the piles varied between 20–80 ft. The natural period also varied. Portions of the freeway built on harder alluvial sediments remained standing.

Although the Cypress structure was built to the standards of its time, it was poorly designed, and it is now recognized that use of nonductile reinforced concrete joints and bents with only three hinges, and inadequate confinement of the structure made its failure predictable.

58 IMPULSE RESPONSE—UNDAMPED SYSTEM

Seismic energy is applied to a structure in a nonregular manner. While a Fourier analysis can be used to analyze the structure response, it is also possible to break the irregular seismic loading into a series of short-duration rectangular impulses. An *impulse* is a force, F, that is applied over a duration, dt, that is much less than the natural period, T, of the structure. The product Fdt is 1.0 for a *unit impulse*. Therefore, a study of the response, $x(t)$, of a system to an impulse is of great interest.[51]

Equation 39 indicates that the same response will be achieved from all short-duration impulses (sine, rectangular, square, triangular, random, etc.) that have the same value of $\int Fdt$. Notice that the response is sinusoidal even though the loading is not.

$$x(t) \approx \frac{1}{m\,\omega} \int (Fdt)\sin \omega t \text{ [undamped]} \quad \text{[SI]} \quad \text{[39(a)]}$$

$$x(t) \approx \frac{g}{W\omega} \int (Fdt)\sin \omega t \text{ [undamped]} \quad \text{[U.S.]} \quad \text{[39(b)]}$$

59 DUHAMEL'S INTEGRAL FOR AN UNDAMPED SYSTEM

If an undamped structure is acted upon by an irregular force of any duration, the loading can be treated as a series of impulses. The response in this case is given by Eq. 40, known as *Duhamel's integral*. Equation 40 is the application of superposition to a series of pulses, each ending at time τ.

$$x(t) = \frac{1}{m\,\omega} \int_0^t F(\tau)\sin \omega(t-\tau)d\tau \quad \text{[SI]} \quad \text{[40(a)]}$$

$$\text{[undamped]}$$

$$x(t) = \frac{g}{W\omega} \int_0^t F(\tau)\sin \omega(t-\tau)d\tau \quad \text{[U.S.]} \quad \text{[40(b)]}$$

$$\text{[undamped]}$$

Several numerical methods can be used to evaluate the integral in Eq. 40. However, when the ground motion is not known in advance, such an integration is not possible. Since earthquake motions are both nonregular and generally unexpected, it is usually acceptable to work

---

[51]An impulse loading can occur from a projectile impact, bomb blast, sudden wind gust, or short-duration seismic tremor.

with a maximum value of acceleration (or velocity or displacement). This is the principle behind the spectral values discussed in Sec. 42. From Eqs. 17 and 18, the total force (i.e., the *base shear*) on the structure is

$$F_{\max} = ma_{\max} \approx \frac{mv_{\max}}{\omega} \qquad \text{[SI]} \qquad [41(a)]$$

$$F_{\max} = \frac{ma_{\max}}{g_c} \approx \frac{mv_{\max}}{\omega g_c} \qquad \text{[U.S.]} \qquad [41(b)]$$

◇   ◇   ◇   ◇   ◇   ◇   ◇

Example 3

A mass of 9.1×10^5 kg (2×10^6 lbm) is supported on two vertical members with lateral stiffnesses of 4.4×10^6 N/m (25,000 lbf/in) each. The columns have no mass and are fixed at both ends. The lateral forcing function consists of a ramp up to 220 kN (50,000 lbf) taking 0.08 s, a uniform loading for 0.08 s, and a ramp down to zero taking 0.08 s. Use Duhamel's integral to determine the response (displacement) as a function of time.

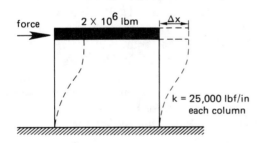

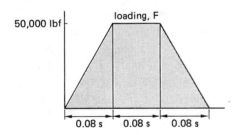

SI Solution

The total combined stiffness, k_t, of the two columns is

$$k_t = (2)\left(4.4 \times 10^6 \frac{\text{N}}{\text{m}}\right) = 8.8 \times 10^6 \text{ N/m}$$

From Eq. 30, the natural period is

$$T = 2\pi\sqrt{\frac{m}{k}} = 2\pi\sqrt{\frac{9.1 \times 10^5 \text{ kg}}{8.8 \times 10^6 \frac{\text{N}}{\text{m}}}} = 2.02 \text{ s}$$

Since the total period over which the loading is applied is much less than the period (3×0.08 s $<$ 2.02 s), the loading can be considered an impulse.

The natural frequency is

$$\omega = \frac{2\pi}{T} = \frac{2\pi}{2.02 \text{ s}} = 3.11 \text{ rad/s}$$

The total impulse is

$$\int F\,dt = \left(\frac{1}{2}\right)(0.08 \text{ s})(220 \times 10^3 \text{ N})$$
$$+ (0.08 \text{ s})(220 \times 10^3 \text{ N})$$
$$+ \left(\frac{1}{2}\right)(0.08 \text{ s})(220 \times 10^3 \text{ N})$$
$$= 3.52 \times 10^4 \text{ N·s}$$

From Duhamel's integral (Eq. 40), the response is

$$x(t) = \int \frac{F\,dt}{m\omega} \sin \omega t$$
$$= \left(\frac{3.52 \times 10^4 \text{ N·s}}{(9.1 \times 10^5 \text{ kg})\left(3.11 \frac{\text{rad}}{\text{s}}\right)}\right) \sin 3.11t$$
$$= (1.24 \times 10^{-2} \text{ m}) \sin 3.11t$$

Customary U.S. Solution

The total combined stiffness, k_t, of the two columns is

$$k_t = (2)\left(25,000 \frac{\text{lbf}}{\text{in}}\right) = 50,000 \text{ lbf/in}$$

From Eq. 30, the natural period is

$$T = 2\pi\sqrt{\frac{m}{g_c k}}$$
$$= 2\pi\sqrt{\frac{2 \times 10^6 \text{ lbm}}{\left(32.2 \frac{\text{ft-lbm}}{\text{lbf-sec}^2}\right)\left(50,000 \frac{\text{lbf}}{\text{in}}\right)\left(12 \frac{\text{in}}{\text{ft}}\right)}}$$
$$= 2.02 \text{ sec}$$

Since the total period over which the loading is applied is much less than the period (3×0.08 sec $<$ 2.02 sec), the loading can be considered an impulse.

The natural frequency is

$$\omega = \frac{2\pi}{T} = \frac{2\pi}{2.02 \text{ sec}} = 3.11 \text{ rad/sec}$$

The total impulse is

$$\int F\,dt = \left(\frac{1}{2}\right)(0.08 \text{ sec})(50{,}000 \text{ lbf})$$

$$+ (0.08 \text{ sec})(50{,}000 \text{ lbf})$$

$$+ \left(\frac{1}{2}\right)(0.08 \text{ sec})(50{,}000 \text{ lbf})$$

$$= 8000 \text{ lbf-sec}$$

From Duhamel's integral (Eq. 40), the response is

$$x(t) = \int \frac{F\,dt}{m\omega} \sin \omega t$$

$$= \left(\frac{8000 \text{ lbf-sec}}{\dfrac{(2 \times 10^6 \text{ lbm})\left(3.11 \dfrac{\text{rad}}{\text{sec}}\right)}{\left(32.2 \dfrac{\text{ft-lbm}}{\text{lbf-sec}^2}\right)\left(12 \dfrac{\text{in}}{\text{ft}}\right)}} \right) \sin 3.11t$$

$$= (0.50 \text{ in}) \sin 3.11t$$

◇   ◇   ◇   ◇   ◇   ◇   ◇

60 MULTIPLE DEGREE OF FREEDOM SYSTEMS

A system with several lumped masses, such as a building with multiple concrete floors supported by steel columns, whose positions are independent of one another is a *multiple-degree-of-freedom* (MDOF) *system*.

An MDOF system has as many ways of oscillating as there are lumped masses. These "ways" are known as *modes*. Each mode has its own characteristic *mode shape* and natural frequency of vibration, each being some multiple of the previous mode's frequency. The mode with the longest period is known as the *first* or *fundamental mode*. Higher modes have higher frequencies (smaller periods), and the periods decrease rapidly

from the fundamental mode.[52] Typical mode shapes of an MDOF system with three lumped masses are shown in Fig. 20.

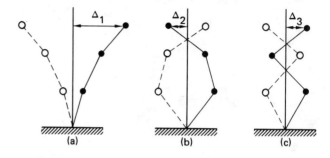

Figure 20   Typical Mode Shapes for a Three-Degree-of-Freedom System

61 RESPONSE OF MDOF SYSTEMS

Each modal frequency results in a specific mode shape, as illustrated in Fig. 20. However, an earthquake contains waveforms with varied frequency content; therefore, all of the modes may be present simultaneously in an earthquake. This makes it difficult to determine the building's response.

Since MDOF response can be determined as the superposition of many SDOF responses, matrix analysis (on a computer) can be used to evaluate MDOF systems based on the equivalent SDOF performance. As with SDOF systems, considerable simplification can be achieved by limiting the analysis to the maximum deflections. However, even this simplification requires a probabilistic analysis because the modal maxima do not occur at the same time, nor do they necessarily have the same sign.

Various approximation formulas are used to combine the modal maxima, and the sum-of-the-squares approximation is commonly quoted.[53] If the maximum displacements, Δ_i, are known for the first n modes for some particular point (e.g., the top story), Eq. 42 usually gives a conservative estimate of the total displacement.

$$\Delta_t = \sqrt{\sum \Delta_i^2} \qquad \text{[42]}$$

[52]For example, for a typical high-rise building with a uniform plan view and a moment-resisting frame, the decrease is in the order of 1, 1/3, 1/5, 1/7, 1/9, and so on.

[53]This is an easy computational approximation. Whether or not it is an accurate approximation is beyond the scope of this book.

The method of combining modal responses by taking the square root of the sum of the squares is referred to in the UBC as the *SRSS method* [Sec. 2337(a)]. An alternative to SRSS is the *Complete Quadratic Combination* (CQC) method described in the App. 1F of the Commentary to the SEAOC Blue Book.

Theoretically, all mode shapes must be included in the summation, but, in practice, most of the vibration energy goes into the first three to six modes, and higher modes can be disregarded. (With the use of a computer, however, there is no need to stop with such a small number of modes.) Since the lower modes dominate, the response spectra for MDOF systems are similar to those of SDOF systems. (See Sec. 65.) For short periods (e.g., less than 1 s), the MDOF response is usually slightly less than for first-mode SDOF systems. For periods exceeding 1 s, the response usually slightly exceeds SDOF response.[54]

The UBC requires that all significant modes be included [Sec. 2335(d)1]. This can be accomplished by making sure that for all modes considered, at least 90% of the mass of the structure is included in the calculation of response for the horizontal direction being investigated [Sec. 2335(e)1].

◇   ◇   ◇   ◇   ◇   ◇   ◇

Example 4

Determine the three modal frequencies for the MDOF system shown.

Solution

Let x_1, x_2, and x_3 be the displacements—measured with respect to the equilibrium position—of masses 1, 2, and 3, respectively. Then, neglecting the inertial (ma) force, the spring forces on each mass are

The free bodies shown are not in equilibrium. (This is particularly evident for mass 3.) According to D'Alembert's *principle of dynamic equilibrium*, an inertial force resisting motion must be added. This inertial force is

$$F_{\text{inertial}} = ma$$

However, from Eq. 18, the acceleration is approximately $\omega^2 x$.

$$F_{\text{inertial}} \approx m\omega^2 x$$

Therefore, the equilibrium equations for the three masses are found by adding the inertial force to the spring forces and then combining coefficients for the three displacements.

mass 1: $\left(m_1\omega^2 - (k_1 + k_2)\right)x_1 + k_2 x_2 = 0$

mass 2: $k_2 x_1 + \left(m_2\omega^2 - (k_2 + k_3)\right)x_2 + k_3 x_3 = 0$

mass 3: $k_3 x_2 + (m_3\omega^2 - k_3)x_3 = 0$

The masses and stiffnesses are known. Writing the three equilibrium equations in matrix form,

$$\begin{bmatrix} \omega^2 - 200 & 100 & 0 \\ 100 & \omega^2 - 200 & 100 \\ 0 & 100 & 0.5\omega^2 - 100 \end{bmatrix} \begin{bmatrix} x_1 \\ x_2 \\ x_3 \end{bmatrix} = \begin{bmatrix} 0 \\ 0 \\ 0 \end{bmatrix}$$

Disregarding the trivial solution, the coefficient matrix must have a determinant of zero. Setting the determinant equation to zero results in the following equation.

$$\omega^6 - (600)\omega^4 + (90,000)\omega^2 - 2,000,000 = 0$$

Being a cubic, this equation has three roots. Each root is a modal frequency.

$$\omega_1 = 5.18 \text{ rad/s}$$
$$\omega_2 = 14.14 \text{ rad/s}$$
$$\omega_3 = 19.32 \text{ rad/s}$$

◇   ◇   ◇   ◇   ◇   ◇   ◇

---

[54]This generalization is highly dependent on the response spectrum and the soil type at the site.

62 MODE SHAPE FACTORS

The *mode shape factors*, ϕ, are relative numbers that represent the ratios of each of the story deflections (from the equilibrium position) to some common basis, usually the deflection of the first or last story. Since mode shape factors are relative, they can usually be determined by initially assuming a value of one of the deflections.

$$\phi_i = \frac{x_i}{x_1} \qquad [43]$$

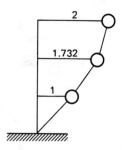

Figure 21   Mode Shape Factors

In some cases, the mode shape factors are normalized by dividing by $\sqrt{\sum m_i \phi_i^2}$. Then Eq. 44 will be valid.

$$\sum m_i \phi_i^2 = 1 \qquad [44]$$

Example 5

Find the normalized first mode shape for the system in Ex. 4.

Solution

The first equilibrium equation is

$$(m_1 \omega^2 - (k_1 + k_2))x_1 + k_2 x_2 = 0$$
$$(1)((5.18)^2 - (100 + 100))x_1 + 100 x_2 = 0$$
$$-173.2 x_1 + 100 x_2 = 0$$

Since the mode shape factors are relative, let $x_1 = 1$. (This will result in an unnormalized mode shape.) Then, $x_2 = 1.732$.

Similarly, the equilibrium equation for mass 3 is

$$k_3 x_2 + (m_3 \omega^2 - k_3)x_3 = 0$$
$$(100)(1.732) + ((0.5)(5.18)^2 - 100)x_3 = 0$$
$$x_3 = 2$$

The unnormalized mode shape is

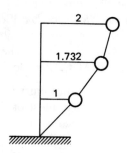

$$\sqrt{\sum m_i \phi_i^2} = \sqrt{(1)(1)^2 + (1)(1.732)^2 + (0.5)(2)^2}$$
$$= \sqrt{6}$$

Dividing each of the unnormalized mode shape factors by $\sqrt{6}$ results in the following mode shape.

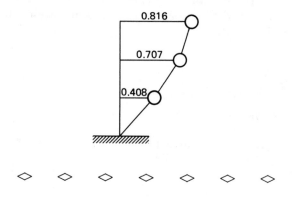

63 RAYLEIGH METHOD

Examples 4 and 5 show the significant computational burden of performing a full dynamic analysis for even a simple MDOF system. While the computer is an ideal tool for doing this, there may be some situations in which such an analysis is unnecessary or inappropriate. (See Sec. 89.)

For such situations, it may be possible to use one of several iterative procedures, most of which are variations of the *Rayleigh method*. This method starts by assuming a mode shape. (Even poor initial assumptions converge rapidly to the correct answer.) Then, the maximum kinetic energy is set equal to the maximum potential energy. Eventually, the mode shape is calculated and used as the starting point for the subsequent iteration.

The *Stodola method*, consisting of the following steps, is one such iterative process.[55]

step 1: Assume a mode shape. That is, assume a deflection, x, for each mass. (A good starting point is the shape taken by the structure when it is turned 90 degrees and acted upon by gravity.)

step 2: Compute the inertial forces for each mass from Eq. 45.

$$F_{\text{inertial}} \approx m\omega^2 x \qquad [45]$$

step 3: Compute the spring forces on each mass as the sum of the inertial forces acting on the springs.

step 4: Compute the spring deflections.

step 5: Calculate the mode deflections from the spring deflections. Repeat from step 1 as required.

◇ ◇ ◇ ◇ ◇ ◇ ◇

Example 6

Use one iteration of the Stodola method to determine the mode shape of the system in Ex. 4.

Solution

step 1: Assume the following mode shape.

$$x_3 = 2.0$$
$$x_2 = 1.5$$
$$x_1 = 1$$
$$x_0 = 0 \text{ (ground)}$$

step 2: The inertial forces are given by Eq. 45.

$$F_{i3} = (0.5)(\omega^2)(2) = \omega^2$$
$$F_{i2} = (1)(\omega^2)(1.5) = 1.5\omega^2$$
$$F_{i1} = (1)(\omega^2)(1) = \omega^2$$

step 3: The spring forces are

$$F_{s3} = F_{i3} = \omega^2$$
$$F_{s2} = F_{i2} + F_{i3} = 2.5\omega^2$$
$$F_{s1} = F_{i1} + F_{i2} + F_{i3} = 3.5\omega^2$$

step 4: The spring deflections are

$$x_{s3} = \frac{F_{s3}}{k_3} = 0.01\omega^2$$
$$x_{s2} = \frac{F_{s2}}{k_2} = 0.025\omega^2$$
$$x_{s1} = \frac{F_{s1}}{k_1} = 0.035\omega^2$$

step 5: Dividing by ω^2, the new relative mode deflections are

$$x_1 = x_{s1} = 0.035$$
$$x_2 = x_{s1} + x_{s2} = 0.060$$
$$x_3 = x_{s1} + x_{s2} + x_{s3} = 0.070$$

Dividing by 0.035, the mode shape is

$$x_3 = 2.00$$
$$x_2 = 1.71$$
$$x_1 = 1$$

These values can be used to repeat the procedure. Eventually, the values from steps 1 and 5 will agree.

◇ ◇ ◇ ◇ ◇ ◇ ◇

64 PARTICIPATION FACTOR

The *participation factor*, Γ_j, is the fraction of the total building mass that acts in any particular mode, j. It can be used to calculate the *story drift*, x. (See also Sec. 63.) The denominator of Eq. 46 is the same as Eq. 44 and will be equal to 1.0 if normalized mode shape factors are used. (If the mode shape factors are normalized, the denominator is not needed.) Weight, W, can be substituted for mass, m, in Eq. 46.

$$\Gamma_j = \frac{\sum m_i \phi_{ij}}{\sum m_i \phi_{ij}^2} \qquad [46]$$

$$x = \Gamma S_d \phi \qquad [47]$$

The participation can also be used to calculate the *floor force*, F_x, that acts at story x (i.e., the force that acts at that level) and the cumulative *story shear*, V_x, that acts at that level and above. This can be done in two ways, one method derived from Hooke's law and using the spring constant, and the other method derived from Newton's law and using the mass. (Section 105 describes the UBC method of distributing the base shear to the stories.)

$$F_x = \Gamma m S_a \phi = \Gamma k S_d \phi \qquad [\text{SI}] \qquad [48(\text{a})]$$

$$F_x = \frac{\Gamma W S_a \phi}{g} = \Gamma k S_d \phi \qquad [\text{U.S.}] \qquad [48(\text{b})]$$

---

[55]The *Holzer method* is another iterative procedure; it is not discussed in this book. See Wakabayashi (1986) and other structural engineering analysis books.

Example 7

Determine the drifts, story shears, and total base shear for the structure in Ex. 5. Assume the spectral displacement and acceleration are 4 (arbitrary units) and 0.28 g (108 in/sec^2), respectively. Assume consistent units are used.

Solution

First, calculate the participation factor from the normalized mode shape factors determined in Ex. 5. Since normalized values are used, the denominator has a value of 1.0 and is not needed.

$$\Gamma = \sum m_i \phi_i$$

$$= (1)(0.408) + (1)(0.707) + (0.5)(0.816)$$

$$= 1.523$$

Use Eq. 47 to calculate the total drifts.

$$x_1 = \Gamma S_d \phi = (1.523)(4)(0.408) = 2.49$$
$$x_2 = (1.523)(4)(0.707) = 4.31$$
$$x_3 = (1.523)(4)(0.816) = 4.97$$

The story drifts are relative to the floors below.

$$x_{3-2} = 4.97 - 4.31 = 0.66$$
$$x_{2-1} = 4.31 - 2.49 = 1.82$$
$$x_{1-\text{ground}} = 2.49 - 0 = 2.49$$

Calculate the story shears from the story drifts. Each of the lateral stiffnesses was 100.

$$V_1 = kx_1 = (100)(2.49) = 249$$
$$V_2 = (100)(1.82) = 182$$
$$V_3 = (100)(0.66) = 66$$

The floor forces can be calculated from the story shears or the participation factors.

$$F_1 = V_1 - V_2 = 249 - 182 = 67$$
$$F_2 = V_2 - V_3 = 182 - 66 = 116$$
$$F_3 = V_3' = 66$$

Alternatively,

$$F_1 = \Gamma m S_a \phi = (1.523)(1)(108)(0.408) = 67.1$$

$$F_2 = (1.523)(1)(108)(0.707) = 116.3$$

$$F_3 = (1.523)(0.5)(108)(0.816) = 67.1$$

The base shear is the sum of the floor forces.

$$V = F_1 + F_2 + F_3$$
$$= 67.1 + 116.3 + 67.1$$
$$= 250.5$$

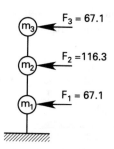

concentrated floor forces

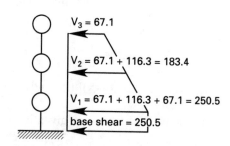

cumulative loading: story shears and base shear

5
RESPONSE OF STRUCTURES

65 ELASTIC RESPONSE SPECTRA

The effective, or spectral, acceleration (Sec. 38) experienced by a building depends on the dynamic characteristics of the building. Specifically, the natural period (Sec. 41) and damping ratio (Sec. 51) are assumed to affect the *spectral acceleration* more than do other factors. For a given damping ratio, ξ, a curve known as a *response spectrum* of spectral acceleration, S_a, can be drawn for various building periods.

There will always be a region on the response spectrum where the acceleration is highest. This occurs where the natural building period coincides with the earthquake period—when the building is in resonance with the earthquake. For California earthquakes, the peak usually occurs in the 0.2 to 0.5 s period range.[56] Theoretically, infinite resonant response (i.e., an infinite magnification factor) is possible, though it is highly unlikely since all real structures are damped.[57]

It seems intuitively logical that a building with large amounts of internal damping will resist acceleration (i.e., motion) to a greater extent than will a similar building with no damping. Such behavior is actually observed as spectral acceleration decreases because damping increases, although the effect of damping at lower periods is slight (since the natural periods of undamped and lightly damped structures are essentially the same). A family of curves (i.e., *response spectra*) for an actual earthquake for various damping ratios is illustrated in Fig. 22. Similar response spectra can be developed for *spectral velocity* and *spectral displacement*.

The spectra shown in Fig. 22 are for *elastic response* to an earthquake. That is, the structures used to develop the curves moved and swayed during the earthquake, but there was no yielding. For that reason, the curves are known as *elastic response spectra*.

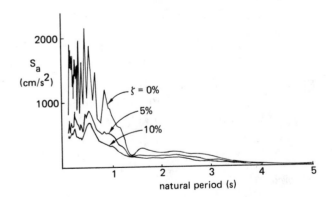

Figure 22   Typical Elastic Response Spectra (1940 El Centro Earthquake in N-S Direction)

Reprinted, with permission, from Minoru Wakabayashi, *Design of Earthquake-Resistant Buildings*, copyright © 1986 by McGraw-Hill Book Company.

---

[56]This is not always the case, as shown by the Loma Prieta earthquake.

[57]A properly designed and constructed building seldom experiences true resonance. Planned or unplanned yielding occurs before true resonant response is achieved, and this yielding damps out the resonance.

66 IDEALIZED RESPONSE SPECTRA

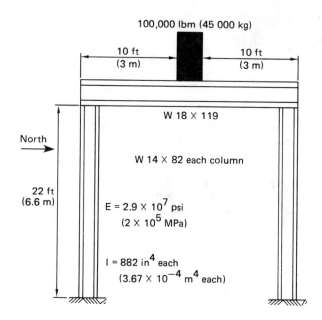

The response spectra derived from the behavior of one SDOF system in one particular earthquake are usually quite jagged, as shown in Fig. 22. It is not possible to use such a historical record for design, since it is unlikely that an earthquake matching the original earthquake in duration, magnitude, or time history will occur. Also, even if the design earthquake was completely specified, the significant variation in spectral values over small period ranges would require an unreasonable accuracy in the determination of the building period. To get around these problems, a smoothed average *design response spectrum* based on the envelopes of performance of several earthquakes is developed.

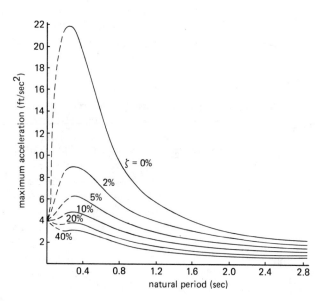

Figure 23   Average Elastic Design Response Spectra
(Based on the 1940 El Centro Earthquake)
[multiplier = 1]

Example 8

The primary support for an industrial drill press with a mass of 45 000 kg (100,000 lbm) is the structural steel bent shown. The beam-column and base connections are rigid. The horizontal beam has a mass of 160 kg/m (119 lbm/ft), neglecting the weight of the vertical supports. The system has 5% damping. Determine the elastic response (i.e., base shear) for a 1940 El Centro earthquake in the north-south direction. Use average design spectra.

SI Solution

The total mass, m, of the moving system is

$$m = 45\,000\,\text{kg} + (6\,\text{m})\left(160\,\frac{\text{kg}}{\text{m}}\right) = 45\,960\,\text{kg}$$

From Table 7, the combined stiffness, k_t, of the two vertical supports is

$$k_t = 2 \times \frac{12EI}{h^3}$$

$$= \frac{(2)(12)(2 \times 10^5\,\text{MPa})\left(10^6\,\frac{\text{Pa}}{\text{MPa}}\right)(3.67 \times 10^{-4}\,\text{m}^4)}{(6.6\,\text{m})^3}$$

$$= 6.13 \times 10^6\,\text{N/m}$$

From Eq. 30, the natural period of vibration, T, is

$$T = 2\pi\sqrt{\frac{m}{k}} = 2\pi\sqrt{\frac{45\,960\,\text{kg}}{6.13 \times 10^6\,\frac{\text{N}}{\text{m}}}}$$

$$= 0.54\,\text{s}$$

From Fig. 23, the spectral acceleration for this period and 5% damping is approximately $S_a = 5.5\,\text{ft/sec}^2$ (1.65 m/s²). From Eq. 17, the base shear is

$$V = mS_a = (45\,960\,\text{kg})\left(1.65\,\frac{\text{m}}{\text{s}^2}\right)$$

$$= 7.58 \times 10^4\,\text{N}$$

Customary U.S. Solution

The total mass, m, of the moving system is

$$m = 100{,}000 \,\text{lbm} + (20 \,\text{ft})\left(119 \,\frac{\text{lbm}}{\text{ft}}\right)$$

$$= 102{,}380 \,\text{lbm}$$

From Table 7, the combined stiffness, k_t, of the two vertical supports is

$$k_t = 2 \times \frac{12EI}{h^3}$$

$$= \frac{(2)(12)\left(2.9 \times 10^7 \,\frac{\text{lbf}}{\text{in}^2}\right)(882 \,\text{in}^4)}{(22 \,\text{ft})^3 \left(12 \,\frac{\text{in}}{\text{ft}}\right)^2}$$

$$= 4 \times 10^5 \,\text{lbf/ft}$$

From Eq. 30, the natural period of vibration, T, is

$$T = 2\pi\sqrt{\frac{m}{g_c k}}$$

$$= 2\pi\sqrt{\frac{102{,}380 \,\text{lbm}}{\left(32.2 \,\frac{\text{ft-lbm}}{\text{lbf-sec}^2}\right)\left(4 \times 10^5 \,\frac{\text{lbf}}{\text{ft}}\right)}}$$

$$= 0.56 \,\text{sec}$$

From Fig. 23, the spectral acceleration for this period and 5% damping is $S_a = 5.5 \,\text{ft/sec}^2$. From Eq. 17, the base shear is

$$V = \frac{mS_a}{g_c} = \frac{(102{,}380 \,\text{lbm})\left(5.5 \,\frac{\text{ft}}{\text{sec}^2}\right)}{32.2 \,\frac{\text{ft-lbm}}{\text{lbf-sec}^2}}$$

$$= 1.75 \times 10^4 \,\text{lbf}$$

◇　◇　◇　◇　◇　◇　◇

67 RESPONSE SPECTRA FOR OTHER EARTHQUAKES

The design response spectra in Fig. 23, although normalized and averaged over several earthquakes, are adjusted for an earthquake of a specific magnitude and peak ground acceleration. Based on historical data and probability studies, the recurrence interval for an earthquake of that magnitude can be determined. For example, an earthquake of the 1940 El Centro magnitude is expected at that site, on the average, every 70 years. However, smaller earthquakes will be experienced more

frequently than every 70 years, and larger earthquakes will be experienced less frequently than every 70 years.

In order to apply the average design response spectra to other earthquakes, they are simply scaled upward or downward for larger and smaller earthquakes, respectively. For example, Table 9 gives the multiplier (to be used with Fig. 23) for other recurrence intervals at the El Centro site.

Table 9
Multipliers for Other Recurrence Intervals
(Based on Elastic Response to the
1940 El Centro Earthquake)

Recurrence Interval (yrs)	Multiplier
2	2.77
20	1.83
32	1.50
70	1.00

68 LOG TRIPARTITE GRAPH

Since spectral acceleration, velocity, and displacement for linear elastic response are all related (see Eq. 18), all three spectral quantities can be shown by a single curve on a graph with three different scales. Such a graph is done on a logarithmic scale and is known as a *log tripartite plot*. Both elastic and inelastic (see Sec. 73) tripartite plots are widely in use. (However, for inelastic response, the spectral acceleration, velocity, and displacement cannot be represented by a single curve on the tripartite plot.)

Tripartite plots, both elastic and inelastic, can differ in how the axes are arranged. Figure 24 illustrates two common arrangements for presenting the information.

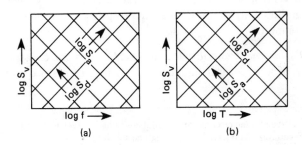

Figure 24　Two Types of Log Tripartite Plots

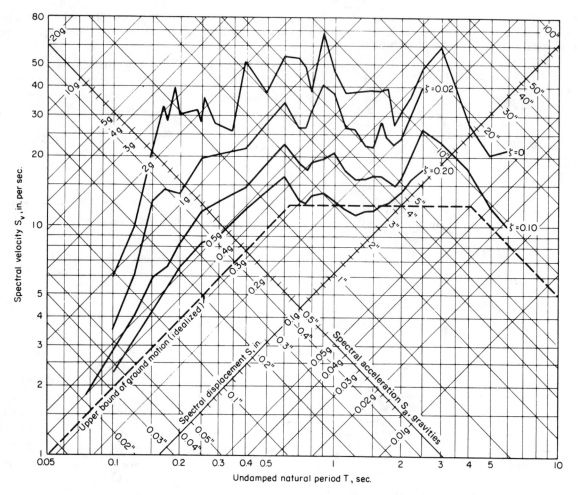

Figure 25 Elastic Log Tripartite Plot (1940 El Centro Earthquake)

Reprinted from *Design of Multistory Reinforced Concrete Buildings for Earthquake Motions*, by John A. Blume,
Nathan M. Newmark, and Leo H. Corning, 1961, with permission from the Portland Cement Association, Skokie, IL.

69 DUCTILITY

The expected magnitude of seismic loads and the nature of building codes make it necessary to accept some yielding during large earthquakes.[58] The design provisions in modern seismic codes could not create a purely elastic response during a large earthquake; in any case, building a structure with such a response would not be economical.

Displacement ductility (or just *ductility*) is the capability of a structural member or building to distort and

yield without collapsing. During an earthquake, a ductile structure can dissipate large amounts of seismic energy even after local yielding of connections, joints, and other members has begun.

The actual ductility of a joint or structural member is specified by its *ductility factor*, μ. There are a number of definitions of the ductility factor, all of which represent the ratio of some property at failure (i.e., fracture) to that same property at yielding. For example, the ductility factor may be specified in terms of energy absorption, as in Eq. 49.

$$\mu = \frac{U_{\text{fracture}}}{U_{\text{yield}}} \qquad [49]$$

In addition to the definition based on the ratio of energies, there are definitions of the ductility factor based on ratios of linear strain and angular strain (rotation). These definitions are not interchangeable, although they

---

[58]There is some circular reasoning here. The high seismic loading expected in California and the high cost of a totally elastic design make it necessary to accept some yielding. Therefore, the building is designed to withstand a smaller effective peak acceleration (see Sec. 38) without yielding, thereby ensuring yielding when a larger ground acceleration is experienced.

are related.[59] Generally, however, the basic concept (i.e., the ratio of some failure property to the same yield property) is all that is needed to explain the significance of a ductile structure.

The minimum ductility of building structures with good connections and good redundancy that are designed to modern seismic codes seems to be about 2.5. (Ductility of bridge structures is much less.) Desirable levels vary, although it is best to have large values of the ductility factor—4 to 6 for concrete frames and 6 to 8 for steel frames.

70 STRAIN ENERGY AND DUCTILITY FACTOR

The area under the stress-strain curve represents the *strain energy* absorbed, U. The maximum energy that can be absorbed without yielding (i.e., the area under the curve up to the yield point) is known as the *modulus of resilience*, U_R. The maximum energy that can be absorbed without failure is the *modulus of toughness (rupture)*, U_T. One definition of the ductility factor, μ, can be calculated from the ratio of these two quantities.

$$\mu = \frac{U_T}{U_R} \qquad [50]$$

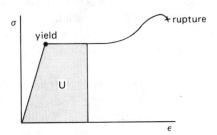

Figure 26　Strain Energy

71 HYSTERESIS

Hysteresis (hysteretic) damping is the dissipation of part of the energy input when a structure is subjected to load reversals in the *inelastic* range. Such dissipation occurs in the structure itself as well as in the soil around the foundation and, therefore, depends on the nature of the building, foundation, and soil. The energy lost per cycle, U_H, is the area within the *hysteresis loop*, as shown in Fig. 27. Hysteresis losses are unaffected by the velocity of the structure.

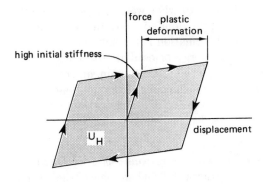

Figure 27　Hysteresis Loop

Inasmuch as it is difficult to evaluate hysteresis losses, hysteresis damping in seismic studies is sometimes accounted for by defining an equivalent internal viscous damping. (See Sec. 50.) Such an approximation works reasonably well in some cases, but the validity deteriorates as the deflection increases.

72 LARGE DUCTILITY SWINGS

The effect of *reversal deformation* after a few cycles of very high ductility is significant. Tests and actual experience indicate that even modern structures can fail after only a few deformation reversals if the strain is well into the inelastic region.[60] It is particularly easy to show that threaded rods are susceptible to such failure.

73 INELASTIC RESPONSE SPECTRA

The total seismic energy, U, received by a building structure in an earthquake is stored or dissipated in four primary ways. Some of the energy is stored as elastic strain energy (U_E); some is converted into kinetic energy (U_K); some is dissipated as hysteretic or plastic losses (U_H); and some is lost due to frictional

---

[59]For ideal (linear) elastoplastic systems, the ductility based on energy absorption, μ_U, can be calculated from the ductility based on strain, μ_ϵ, as

$$\mu_U = 2\mu_\epsilon - 1$$

This means that if the ductility, as calculated from linear strain, is 4 to 6, the ductility will be 7 to 11 when calculated from Eq. 49.

---

[60]This type of failure was first predicted in a controversial paper written by Vitelmo V. Bertero and Egor Popov in the 1960s. Such failures were actually observed in the 1964 Alaska and 1971 San Fernando earthquakes.

and damping effects (U_ξ). In the simplest models, the sum of these four terms equals the total energy input.

Particularly when the building is stressed in the elastic region, input energy is dissipated relatively slowly, primarily because of internal friction (i.e., damping) of the structure converting the kinetic energy into heat. However, it takes much more energy to plastically deform parts of the structure. Since the amount of input energy is limited to what was received and the frictional losses are approximately constant, an increase in this energy of deformation is accompanied by a decrease in the kinetic energy of oscillation. Thus, each yielding connection, every broken column, and every sheared pin dissipates a finite amount of kinetic energy. Therefore, a building's amplitude of oscillation and number of oscillation cycles decrease as major portions of the building yield.[61]

The *inelastic design response spectra* (IDRS) show what the acceleration will be when some of the seismic energy is removed inelastically. It is appropriate to consider the inelastic effects when the response of a building to a major earthquake is being determined. The inelastic response spectra are usually derived from the elastic response spectra.

There are several well-known methods of obtaining the inelastic response spectra from the elastic response spectra, but few of them are suitable for manual calculations. Perhaps the quickest and easiest, though not necessarily the most rigorous, method is simply to scale the elastic curves downward by some function of the ductility factor.

$$S_{a,\,\text{inelastic}} = \frac{S_{a,\,\text{elastic}}}{\text{factor}} \qquad [51]$$

The "factor" in Eq. 51 depends on the period. For extremely small periods (i.e., frequencies greater than approximately 33 Hz), there is no reduction at all. For periods greater than approximately 0.5 s to 1.5 s (i.e., frequencies less than 2 Hz), the ductility factor (based on strain), μ_ϵ, itself can be used as the reduction factor. For intermediate periods (33 Hz > f > 2 Hz), the reduction factor is approximately $\sqrt{\mu_U} = \sqrt{2\mu_\epsilon - 1}$.

In converting an elastic response spectrum to an inelastic response spectrum, the ductility factor, μ_ϵ, used to calculate the reduction factor may be known as the *structure deflection ductility factor* or *design ductility factor*, μ_Δ. It is the ratio of the deflection at ultimate collapse to the deflection at first yield, measured at the roof of the structure. Estimates of this value are known

to be unreliable at low natural periods (i.e., high frequencies), but the simple division by μ_Δ or μ_ϵ is favored because of its simplicity.

Values of the design ductility factor in excess of 6 are not often used, as excessive damage (architectural as well as structural) would be experienced, even though larger values (up to 10 for ductile steel structures) are readily achievable.

At high periods (i.e., low frequencies), energy absorption effects dominate, and a ductility factor based on energy (rather than strain), μ_U, is more appropriate for use in determining the inelastic response spectrum. (See Ftn. 59.)

74 DRIFT

Drift (also known as *story drift*) is the deflection of one floor relative to the floor below. There are two main reasons to control drift. First, excessive drift in upper stories has strong adverse psychological and physical effects on occupants. Second, it is difficult to ensure structural and architectural integrity with large amounts of drift.[62] Excessive drift can be accompanied by large secondary bending moments and inelastic behavior. (See Sec. 75.) In a severe earthquake in which yielding is experienced, a modern high-rise building can be expected to experience a drift of approximately 1.5% of its total height at the uppermost story.[63]

There are three components of drift: (1) column bending from lateral and eccentric loads, (2) joint rotation due to transverse loads, and (3) frame bending from axial loads. The first two components together are referred to as *shear drift*. The third component is known as *chord drift*.

---

[61] A yielding structure experiences larger localized deformations than an elastic structure does. This is different, however, from the overall oscillation of the structure.

[62] *Architectural failures* are such nonstructural damage as failure of partitions, windows, and hung ceilings. In low-rise construction, damage to stairwells and elevator shafts can also be considered nonstructural. However, in high-rise construction, stairwells and elevator shafts usually constitute the most critical structural elements in the structural core.

[63] The UBC limits the drift under the code-specified design lateral forces to much less than this—approximately 0.5% of the height [UBC Sec. 2334(h)2]. (See Sec. 107.) Under larger forces, the drift will be larger.

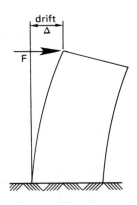

Figure 28 Drift

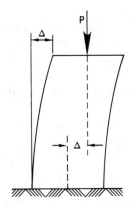

Figure 29 P-Δ Effect

Table 10 contains generalizations about the effects of different variables on drift. The drift is proportional to the variables (raised to the powers indicated) defined in the table.

Table 10

Effect of Different Variables on Drift

Variable	Column Drift	Girder Drift	Joint Drift	Chord Drift
story height, H	H^3	H^2	H^2	none
building height, H	none	none	none	H^3
girder length, L	none	L	none	L^3
column depth, D	D^2	none	D^{-1}	none
girder depth, D	none	D^2	D^{-1}	none
column height, H	H	none	H^2	none
shear load, F	none	none	none	F
frame length, L	none	none	none	L^{-2}

The *story drift ratio* is the story drift divided by the height (floor to floor) of the story.

75 P-Δ EFFECT

The column members in a building are loaded in compression by the vertical live and dead loads. Normally, these loads are concentric with the base of the building. When the building is acted upon by a lateral (horizontal) seismic load, however, the vertical loads are eccentric with respect to the base. The *overturning moment* adds an eccentric bending stress to the columns. The magnitude of the overturning moment is $P\Delta$, where P is a function of the building weight and Δ is the drift. The additional column stress is referred to as the P-Δ *effect*.

If the overturning moment increases faster than the restoring moment from the frame stiffness, the frame is unstable. Since the vertical load is constant (i.e., is not transient like the seismic load that causes the initial eccentricity), the column members will eventually fail and the frame will buckle. Protection against instability failures is provided by wall X-bracing and thick shear walls.

Unstable frames can be inadvertently designed in nonseismic areas where only vertical loads are considered. Designing a frame to withstand large lateral (seismic) loads has the effect of limiting drift; such frames, therefore, are unlikely to experience a problem caused by P-Δ instability.

76 TORSIONAL SHEAR STRESS

A building's *center of mass* (on plan view) is a point through which the base shear (i.e., the total lateral seismic force) can be assumed to act. This base shear is resisted by the vertical members at the ground level. Each such member may have a different rigidity and thus provides a different lateral resisting force in the opposite direction of the base shear. The building's *center of rigidity* is a point through which the resultant of all the resisting forces acts.[64]

If the building's center of mass does not coincide with its center of rigidity, the building will tend to act as if it is "pinned" at its center of rigidity. It is said to

---

[64]The implication here is that the structure has rigid diaphragms (see Sec. 118) between the floors so that the torsional moment can be transferred to the various resisting members distributed at that level. Structures with flexible diaphragms (see Sec. 120) are incapable of distributing torsional moments to vertical resisting elements. The UBC [Sec. 2334(f)] gives a method of determining whether or not a diaphragm can be considered flexible. Specifically, the diaphragm is flexible if the maximum lateral diaphragm deformation is more than twice the average story drift.

be acted upon by a torsional moment, $M_{torsional}$, calculated as the product of the base shear, V, and the eccentricity, e. This *eccentricity* is the distance (measured perpendicular to the base shear's line of action) between the centers of mass and rigidity.

$$M_{torsional} = Ve \qquad [52]$$

The rotation about the center of rigidity is resisted by a torsional shear stress in all members.[65] This torsional stress is proportional to the distance, r, from the building's center of rigidity to the resisting member. In Eq. 53, J is the polar moment of inertia of the resisting members and is most conveniently calculated as the sum of the moments of inertia taken with respect to the x- and y-axes.

$$v = \frac{M_{torsional}r}{J} \qquad [53]$$

$$J = I_x + I_y \qquad [54]$$

The polar moment of inertia, J, can be calculated from the relative rigidities of the resisting elements. If R_i is the relative rigidity of shear wall i and r_i is the distance of wall i from the center of rigidity, then the polar moment of inertia is

$$J = \sum R_i r_i^2 \qquad [55]$$

The units of J are somewhat ambiguous since the units used to determine the relative rigidities are not specifically known. However, the units are definitely not the usual $(area)^4$.

The shear force, $F_{i,torsion}$, in member i due to torsion is

$$F_{i,torsion} = \frac{R_i r_i M_{torsional}}{J}$$
$$= \frac{R_i r_i M_{torsional}}{\sum R_i r_i^2} \qquad [56]$$

Then, the torsional shear stress in member i is found by dividing the shear force by the cross-sectional area of the member.

$$v = \frac{F_{i,torsion}}{A} \qquad [57]$$

The contribution of a stiff element to torsional rigidity increases with the square of the distance of that element from the center of rigidity. If R is the relative rigidity of a shear wall and r is the distance of the wall from

the center of rigidity, then the contribution of the wall to the torsional moment of inertia is $J_w = Rr^2$. Therefore, shear walls should be located as near the building perimeter (and hence as far from the center of rigidity) as possible.

The UBC [Sec. 2334(e)] requires that an *accidental eccentricity* of 5% (based on the maximum building dimension at that level perpendicular to the seismic load) be added to the actual eccentricity, if any, in the design of all buildings, even those that are symmetrical. (Also, see Sec. 106.) This eccentricity is included to account for accidental errors in workmanship, uncertainties in the actual location of the centers of mass and rigidity, nonuniform distribution of dead and live loads, nonuniformities that result from subsequent building modifications, and eccentricities that develop during an earthquake after the failure of certain structural elements.

Example 9

A crane system is modeled as a 455-kg (1000-lbm) mass attached to the end of a 1.5-m (5-ft) cantilever beam supported by a 51-cm (20-in) diameter hollow tubular column. Calculate the maximum torsional shear stress for a lateral acceleration in the x-direction of 0.3 g.

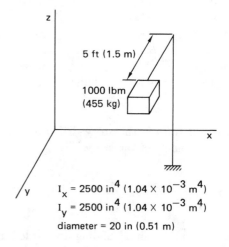

$I_x = 2500$ in^4 (1.04×10^{-3} m^4)

$I_y = 2500$ in^4 (1.04×10^{-3} m^4)

diameter = 20 in (0.51 m)

SI Solution

The lateral (seismic) force is equal to the inertial force.

$$F_i = ma = (455\text{ kg})(0.3\text{ g})\left(9.81\ \frac{\text{m}}{\text{s}^2 \cdot \text{g}}\right)$$
$$= 1339\text{ N}$$

---

[65]Unlike the base shear, which is resisted only by walls parallel to the seismic force, the torsional shear is resisted by all walls and columns.

The torsional moment is

$$M_{\text{torsional}} = Fe = (1339 \text{ N})(1.5 \text{ m})$$
$$= 2009 \text{ N·m}$$

Equation 54 gives the polar moment of inertia.

$$J = I_x + I_y$$
$$= 1.04 \times 10^{-3} \text{ m}^4 + 1.04 \times 10^{-3} \text{ m}^4$$
$$= 2.08 \times 10^{-3} \text{ m}^4$$

The distance from the center of rigidity (i.e., the center of the column) to the most exterior point on the column is 25.5 cm. The maximum torsional shear stress is given by Eq. 57.

$$v = \frac{Mr}{J}$$
$$= \frac{(2009 \text{ N·m})(0.255 \text{ m})}{2.08 \times 10^{-3} \text{ m}^4}$$
$$= 2.46 \times 10^5 \text{ Pa}$$

Customary U.S. Solution

The lateral (seismic) force is equal to the inertial force.

$$F_i = \frac{ma}{g_c} = \frac{(1000 \text{ lbm})(0.3 \text{ g})\left(32.2 \frac{\text{ft}}{\text{sec}^2\text{-g}}\right)}{32.2 \frac{\text{ft-lbm}}{\text{lbf-sec}^2}}$$
$$= 300 \text{ lbf}$$

The torsional moment is

$$M_{\text{torsional}} = Fe = (300 \text{ lbf})(5 \text{ ft})\left(12 \frac{\text{in}}{\text{ft}}\right)$$
$$= 18{,}000 \text{ in-lbf}$$

Equation 54 gives the polar moment of inertia.

$$J = I_x + I_y = 2500 \text{ in}^4 + 2500 \text{ in}^4$$
$$= 5000 \text{ in}^4$$

The distance from the center of rigidity (i.e., the center of the column) to the most exterior point on the column is 10 in. The maximum torsional shear stress is given by Eq. 53.

$$v = \frac{Mr}{J} = \frac{(18{,}000 \text{ in-lbf})(10 \text{ in})}{5000 \text{ in}^4}$$
$$= 36 \text{ lbf/in}^2 \text{ (psi)}$$

◇   ◇   ◇   ◇   ◇   ◇   ◇

77 NEGATIVE TORSIONAL SHEAR STRESS

The base shear causes a shear stress that acts in the same direction in all vertical base members. The torsional shear stress, however, has different signs on either side of the center of rigidity. On one side (i.e., where the resisting element is on the same side of the center of rigidity as the center of mass) the torsion increases the stress from the base shear; on the other side, the stress is decreased. The amount of decrease is known as *negative torsional shear stress*. Negative torsional shear stress should normally be neglected; that is, it should not be used to decrease the design capacity of a wall or member.

It is easy to make the error of reversing the signs of the induced stresses and adding the negative torsional stress where it should be subtracted, and vice versa. The key to avoiding this error is always to work with the stresses that *resist* the forces and moments. Thus, the stress that resists the base shear acts in a direction opposite to the base shear (i.e., opposite to the direction of ground motion). Similarly, the torsional stresses that resist the torsional moment are in the direction opposite to the applied moment.

78 OVERTURNING MOMENT

The summation of all moments taken about the base due to the distributed lateral forces (see Sec. 105) is the *overturning moment*, often given the symbol OTM. If the overturning moment is large enough, it can reverse the compression that normally exists in outer columns caused by the dead and live building loads. Because footings and concrete walls and columns can be placed in a state of tension, the overturning moment is more of a problem for concrete frame and shear wall construction (which cannot tolerate much tension) than it is for steel frame construction.

The overturning moment will increase the compressive stress in outer columns on the opposite side of the building. Such an increase must be countered by increasing the thickness of shear walls and using extra steel reinforcement in concrete columns.

Overturning moments should be calculated for each building level. The first overturning moment is the sum of all moments taken about the ground level. This moment should be used to size footings and to design the primary outer columns. The overturning moment for each subsequent floor considers only lateral forces above that floor. This moment is used to design the shear walls and other supporting structures at that floor.

79 RIGID FRAME BUILDINGS

Before 1965, when the design of structural systems was still in its infancy, most tall buildings were designed as *rigid frames*.[66] In a rigid-frame building, columns and beams were welded together to create a structural grid that resisted wind and earthquake forces elastically.[67] Such buildings were expensive to construct because they used inordinately large amounts of material, usually steel, to keep the stresses in the elastic region.[68]

80 HIGH-RISE BUILDINGS

The optimum design for seismic loading often conflicts with that for wind loading, a significant factor for any tall building. For an earthquake, the building needs to be flexible, even though the full flexibility might be called upon only once in 400 years. However, the full flexibility might be experienced during large windstorms, say, once every ten years. The greater flexibility required to resist large earthquakes makes for unpleasant motions in windstorms.[69]

The rigid-frame system relies on the bending of columns and beams for its lateral stiffness. However, bending is a poor way to tap a structural member's strength compared to axial loading.[70] A *tube building* resists lateral forces in a radically different way from a rigid-frame building. The tube is like a giant box beam cantilevering out of the ground. Axial forces in the columns mainly resist the tendency to move laterally.

In order to economically design for increasing numbers of stories, different flexible structural systems were developed.[71] The most general systems are: (1) frames with bracing in the core, which creates a stiff vertical truss, good for buildings up to 30 or 40 stories, (2) framed tubes, good for up to 60 or 70 stores, and (3) diagonally-braced tubes, good for up to 100 or 120 stories.

In a *pure* tube system, all of the lateral resistance is in the structure's exterior tube, made up of closely-spaced columns linked by stiff spandrel beams. It is also possible to have *tube-in-tube* systems in which additional columns or shear walls closer to the center of the building, such as those surrounding an elevator shaft, provide additional axial load resistance. Tube-in-tube systems seem to be popular in *composite buildings*, which use concrete (or concrete-encased steel) columns for the outer tube and steel columns for the inner tube.

---

[66]The Empire State and the Chase Manhattan Bank buildings in New York City and the Tenneco Building in Houston are rigid-frame structures.

[67]Bracing in the core (see Sec. 80) was not used, although it was recognized that it contributed to structural performance. Using such bracing would have been prohibitively complicated because tools for the structural analysis, such as computers and software, had not been developed. The increasing cost of land after 1965 also made it worthwhile to use costlier designs.

[68]For smaller buildings, rigid frames may be more economical in some cases, particularly where wind forces prevail or when the cost of the additional material is less than the cost of increased design and testing.

[69]This and the preceding sections are meant to document the trend in high-rise design, not to suggest using tube structures in designs for earthquake resistance.

[70]A measure of the "efficiency" of steel is the weight of steel per square foot of floor space for all stories. The 60-story rigid-frame Chase Manhattan Bank Building uses about 60 pounds of steel per square foot ($290 \ kg/m^2$). The 100-story John Hancock Center in Chicago uses a trussed tube structural system requiring half as much steel per unit area—about 30 pounds per square foot ($145 \ kg/m^2$).

[71]Dr. Fazlur Khan is acknowledged as being the first structural engineer to recognize the value of the alternate structural systems. One of the first (if not the first) flexible buildings was the 43-story concrete framed-tube system Chesternut Dewitt Apartment Building in Chicago that Khan designed. He also designed the One Shell Plaza building in Houston (a 50-story structure using lightweight concrete).

6

UBC SEISMIC CODE

81 HISTORICAL BASIS OF SEISMIC CODES IN CALIFORNIA

It is somewhat surprising that the formal study of earthquake-resistant design had to wait until considerably after the 1906 *San Francisco earthquake*. The only lateral force requirement placed on structures designed and constructed in San Francisco after that earthquake was a 30 lbf/ft^2 wind loading.

Only after the 1925 *Santa Barbara earthquake* did the California legislature direct that significant effort be expended on the study of seismology.[72]

There were 115 fatalities in the 1933 *Long Beach earthquake*. Inasmuch as there was widespread damage (caused by poor workmanship, design errors, and construction shortcuts) to schools in the area, the *Field Act* was subsequently passed. This act gave the Division of Architecture, State Department of Public Works, responsibility for approving school designs.[73] Also, the

1933 Riley Act set minimum standards for lateral force resistance in all buildings (specifically, just 2% of the dead load). The format of the early codes was that the building had to be "strong" enough to resist a static lateral force, the *base shear*, V, of some fraction (e.g., 10% for masonry school buildings) of the weight, W.[74] The fraction was known as the *base shear coefficient*, C. Between 1943 and 1953, the base shear coefficient was modified several times based on the building period and/or the height of a building, but the *equivalent static force* concept remained and is in use to this day.[75,76]

$$V = CW \qquad [58]$$

Nine lives were lost in the 1940 *El Centro earthquake*, a 7.1 magnitude event caused by the *Imperial Fault*. While only approximately 10 s in duration, a relatively high ground acceleration, 0.33 g, was observed. This earthquake was significant, not because of the widespread damage, but because it was the first earthquake

---

[72]Based on damage reports, the 1925 Santa Barbara earthquake is estimated at 6.3 Richter magnitude, although Richter-style seismometers had not yet been developed. Fatalities were limited, primarily because the earthquake occurred in the early morning before people were in the business district and children were in school. Widespread damage similar to that which destroyed San Francisco was averted as city engineers detected the tremors by observing fluctuations in the water pressure gauges and shut off the gas valves and electrical mains. Despite this, there were 13 fatalities and significant building damage, particularly in brick, masonry, and tile construction. Steel, wood, and properly designed reinforced concrete construction sustained little or no damage, although damage to poorly designed concrete structures occurred.

[73]The 1933 Long Beach earthquake had a Richter magnitude of 6.3, according to the newly-developed seismometer. (Although the seismometer's range was exceeded, there is essentially a full record of this earthquake.) As with the 1925 Santa Barbara earthquake, the 1933 Long Beach earthquake occurred in the

early morning, before children were in school. The same types of structural failures were observed—that is, masonry and brick buildings, in particular, were damaged.

[74]This is simplifying the theory slightly, as the base shear coefficient was actually applied to the dead load and some part of the live load.

[75]The American Society of Civil Engineers (ASCE) and the Structural Engineers Association of Northern California (SEAONC) formed a committee in 1948 that recommended that the equivalent static force concept be used in San Francisco. In 1959, the Structural Engineers Association of California (SEAOC) code was expanded to a "uniform" code for all areas of the United States. This was the first "Blue Book." At this time, the type of building—frame, box, and so on—was made significant in code specifications.

[76]There are cases in which the UBC [Sec. 2333(h)3] requires a dynamic analysis. (See Sec. 89.)

to occur in a heavily-instrumented area. The first accelerometer yielding response data on building periods was obtained.

The 1966 *Parkfield earthquake* on the San Andreas Fault had a relatively low magnitude of 5.5 and a very short duration, but the ground acceleration of 0.5 g was the highest observed to that date. Thus, it is apparent that magnitude and acceleration are not necessarily correlated.

A great amount of accelerometer data was obtained from the 1971 *San Fernando earthquake* (6.6 magnitude, San Fernando Fault zone). This earthquake was significant for several reasons. First, an unbelievable ground acceleration of 1.24 g was experienced at the Pacoima Dam site. Second, there were failures of new buildings designed with the current seismic codes.[77]

The 1979 *Imperial Valley earthquake* (Richter magnitude 6.6) produced the first accelerometer data from a building with extensive damage. In a building that was partially supported at the ground level by concrete columns, the period and amplitude of oscillation decreased significantly each time one of the columns failed. This is consistent with the concept that seismic energy is removed from a yielding structure.[78]

The 1989 *Loma Prieta earthquake* (magnitude 7.1, 62 fatalities, San Andreas Fault) was significant because of the important lessons learned about soil and site conditions. The most significant damage outside of the epicenter occurred in areas where soil resonance magnified the seismic energy. Although the overall seismic energy should have been (and probably was) greatly attenuated by the large distance between the epicenter and San Francisco, the site conditions under the San Francisco Marina district and Oakland's Interstate 880 Cypress structure magnified the remaining energy.

82 SEISMIC CODES

A *code* is a set of rules adopted by an organization empowered to enforce the code. The mere publication of a set of guidelines such as contained in the SEAOC Blue Book does not constitute a governing law. The guidelines must be adopted by a law-making body to become legal documents.

---

[77]In particular, the New Olive View Hospital and the San Fernando Veterans Administration Hospital were new structures that sustained major damage.

[78]That this event proves inelastic behavior removes seismic energy from a structure should not be used to legitimize intentional design for inelastic behavior. It would be a major flaw to design columns to lose their load-carrying ability in the way they did in this instance.

The *Uniform Building Code (UBC)* contains the most extensive seismic provisions for buildings of any code.[79] It is derived from the *Recommended Lateral Force Requirements* (commonly referred to as the *Blue Book*) published by the Seismology Committee of the Structural Engineers Association of California, SEAOC.[80] The Blue Book Commentary is not reproduced in the UBC, but this commentary is invaluable in understanding the significance of the UBC code provisions.

The SEAOC Blue Book provisions have been included in the UBC since approximately 1960, but other building codes were slower to include more than limited seismic provisions, probably because the true seismic risk of the regions that have adopted those codes was not recognized.[81] However, following the 1971 San Fernando earthquake, when several buildings supposedly built according to current seismic provisions experienced substantial damage, other organizations began writing seismic provisions.

⋄ The Applied Technology Council (ATC) published ATC 3-06 in 1978. This was a massive 500-page document intended to serve as a reference for other code-writers. It has now been superceded by the NEHRP provisions.

⋄ The Building Seismic Safety Council (BSSC), with Federal Emergency Management Agency (FEMA) funding, published the National Earthquake Hazards Reduction Program (NEHRP) provisions in 1985 and revised them in 1988.

⋄ The American National Standards Institute (ANSI) published its A58.1 in 1982. This document deals with determining seismic loading, but it does not address detailing.

⋄ The American Concrete Institute (ACI) included detailing to resist seismic loads as Appendix A in the 1983 edition of ACI-318. However, determination of seismic loading is not covered.

⋄ The American Institute of Steel Construction has begun work on a document providing detailing requirements for steel buildings. As with the ACI document, determining the seismic loads is not covered.

---

[79]Published by the International Conference of Building Officials in Whittier, California.

[80]SEAOC, 217 Second Street, San Francisco, CA 94105.

[81]The nation's two other model building codes are written by the Building Officials and Code Administrators International (BOCA) in Country Club Hills, Illinois, and the Southern Building Code Congress International (SBCCI) in Birmingham, Alabama. Both codes significantly strengthened their seismic provisions in 1992. However, the two codes differ from the UBC in their methodology.

While the large number of seismic documents may seem confusing, it should be noted that all were used as "source documents" for the 1988 (and subsequent) Blue Book and 1988 (and subsequent) UBC. In fact, the 1988 Blue Book drew its format for the base shear equation from the ATC document.[82]

Adoption of the UBC is up to each municipality. Most large cities have their own specific requirements that can supersede portions of the UBC or replace it entirely. Design of high-rise buildings located in Los Angeles, for example, is governed by a city code, not the UBC.

General seismic provisions such as those in the UBC may be superceded by even more stringent statutory requirements. For example, Title 24 of the California Administrative Code requires that schools and hospitals be operational after an earthquake.

Bridges in California are designed according to CALTRANS/AASHTO seismic provisions. These are similar in concept to the UBC seismic provisions. However, the nomenclature is different, and the response spectra provided are more detailed.

The CALTRANS/AASHTO method evaluates each bridge two ways (the term "loading" refers to lateral loads, moments, and shears):

1. transverse seismic loading plus 30% of the longitudinal seismic loading

2. longitudinal seismic loading plus 30% of the transverse seismic loading

As with the UBC method, the CALTRANS/AASHTO method includes both static and dynamic analyses. The static method is used for well-balanced spans with supports that are all approximately equal in stiffness. Once determined, the maximum lateral load is applied uniformly along the bridge. The dynamic method is used when the bridge is irregular in configuration or strength.

With the static method, the earthquake design force, V, is calculated as

$$V = \frac{(\text{ARS})W}{Z} \qquad [59]$$

ARS is the Acceleration Response Spectrum value obtained from one of four response spectra curves provided in the code. Different curves are provided for different alluvium (i.e, soil) depths and peak rock accelerations for the *maximum credible earthquake* expected in that area. (The maximum credible earthquake is obtained from maps published by the California Division of Mines and Geology.)

Z is the ductility risk reduction factor which depends on the type of structure and its period. Z varies from slightly less than 1.0 to slightly more than 8.0.

W is the dead load (i.e., the weight) of the bridge.

The design bridge period, T, (in seconds) is calculated from the following formula. P is the stiffness of the superstructure (i.e., the total force that, if applied uniformly, would cause the bridge to deflect one inch). Consistent units must be used.

$$T = 0.32\sqrt{\frac{W}{P}} \qquad [60]$$

83 THE SURVIVABILITY DESIGN CRITERIA

While little effort is expended in trying to design buildings that will be totally elastic (i.e., experience no damage) during an earthquake, it is implicit in seismic codes that catastrophic collapse must be avoided. The following three design standards constitute the implied UBC seismic *survivability* (or *life-safety*) *design criteria* [1990 Blue Book Commentary, Sec. 1A.1]. It is notable, however, that these criteria are not actually specified in the UBC.

◇ *There should be no damage to buildings from a small earthquake.*

◇ *There may be minor architectural damage, but no structural damage during a moderate earthquake.*

◇ *There may be structural damage but no collapse during a severe earthquake.*[83] Yielding is relied upon to dissipate the damaging seismic energy. Theoretically, all structural damage will be repairable when collapse is prevented, although some buildings may be condemned and replaced for reasons of economics or convenience.

---

[82]There was no "rational" basis for the individual parameters in the earlier base shear equation. In the new code, all of the factors in the numerator logically represent characteristics that contribute to increased loading. (The numerator reduces the actual numerical value to a "consensus value" of design base shear.)

[83]These italicized sentences are the standards of survivability as they are commonly stated, but everyone knows that the degree of damage is dependent on the severity of the ground shaking at the building site, not on the magnitude of the earthquake at some distant epicenter.

84 EFFECTIVENESS OF SEISMIC PROVISIONS

The code provides "reasonable" but not complete assurance of the protection of life. Furthermore, the code does nothing to prevent construction on land that is subject to earth slides (of the type that occurred during the Alaskan earthquake) or liquefaction (as occurred in the Niigata, Japan earthquake).

It is important to note that the UBC provisions are intended as minimum requirements. The level of protection can be increased by increasing the design lateral force, energy absorbing capacity, redundancy, and construction quality assurance.

It is also important to note that seismic design is not a science—it is an art that, unfortunately, must be verified in the field. Thus, the nature of a seismic code is to require design features or methods and then observe the effectiveness of those features.

Finally, the seismic code used is not the only factor affecting the performance of a structure during an earthquake. In most cases of structural failure in modern buildings, earthquake severity, duration, soil conditions, inadequate design, poor control or material quality, and poor workmanship are found to be the major factors contributing to collapse.[84] Of course, modern seismic codes cannot be blamed when pre-1976 structures fail. (See Sec. 34.)

85 APPLICABLE SEISMIC SECTIONS IN THE UBC

The major seismic provisions are contained in Chapter 23, Part III of the UBC. However, the code provisions for sizing and detailing structures appear in other chapters. Specifically, Chap. 26 (in particular, Sec. 2625) for concrete, Chap. 27 (in particular, Sec. 2710) for steel, Chap. 25 for timber, and Chap. 24 (in particular, Sec. 2407(h)) for masonry contain these provisions.[85]

---

[84]In particular, the widespread structural failures that occurred in the 1985 Mexico City earthquake are examples of how even modern buildings can be "brought down" by these contributing factors. In fact, the failures that occurred seem to validate the need for the current UBC provisions, as the very features required by that code were often not included in the design of buildings that collapsed.

[85]It is important to recognize that the provisions of the UBC Chap. 26 are not, by themselves, usually sufficient to design reinforced concrete building structures in California, as the provisions deal with ordinary concrete. These sections must be taken in conjunction with Sec. 2625, which details the requirements for *specially-confined concrete*. There are exceptions [UBC

Although a section for nonbuilding structures was added in the 1980 version, Chap. 23 of the UBC primarily covers buildings, but not such structures as retaining walls, most towers and tanks, bridges, dams, docks, and offshore platforms.

Other parts of the UBC that are occasionally useful are Tables 25-G and 25-Q (nailing), Table 25-F (bolting in wood), and Table 24-D-1 (straps and ties in masonry).

86 THE NATURE OF UBC SEISMIC CODE PROVISIONS

There are two major categories of seismic provisions in the UBC: (1) those that relate to proportioning the structure, and (2) those that relate to detailing elements of the structure. The proportions are chosen such that the structure's ability to absorb energy (i.e., its "strength") matches the application of energy, no matter how much yielding has occurred, and such that overall stability is maintained. This requires that the lateral force resisting elements be roughly distributed (in plan) throughout the structure. (Thus, arbitrarily increasing the strength of one element may actually have a negative effect on the overall seismic performance.) In equation form, the ratio of *energy demand* to *energy capacity* evaluated in plan should be roughly constant.

Design details prevent premature local failure by ensuring ductile behavior and preventing local instability and failure of elements that are cyclically stressed beyond their yield points. Unlike the UBC provisions for proportioning the structure, the design details can usually be determined without evaluating the stresses, drift, or loads.

Controlled yielding in a major earthquake is implicitly anticipated by the UBC, and, therefore, a code based on yield or ultimate strengths would be preferred. However, the seismic design forces for steel design in the UBC are based on an *allowable stress design (ASD)* (*working stress* or a *service level stress*) and not a yield stress basis. This is primarily due to the fact that the steel structural systems used in the majority of high-rise structures are still, for the most part, based on ASD.[86]

The American Institute of Steel Construction (AISC) has published its own design standards (which compete

---

Sec. 2625(c)1F], but the burden of proof when applying the exceptions is on the designer.

[86]While concrete is designed almost totally with ultimate strength models, the *load resistance factor design (LRFD)* methods available for steel have not been widely accepted by American engineers. Sufficiently detailed factored load provisions also have yet to be adopted for timber and masonry construction.

with the UBC) for steel buildings in seismic zones. However, these standards apply only to buildings designed by the load resistance factor design (LRFD) method. The new standards do not apply to buildings designed by the allowable stress design (ASD). The standards cover the design of beams, columns, concentric- and eccentric-braced frames, and moment-resisting connections, including panel zones at the intersections of beams and columns.

87 ALLOWABLE STRESS LEVELS

When the *allowable stress design criterion* is used in the design of steel, timber, and masonry structures, the UBC [Sec. 2303(d)] permits a one-third increase in allowable stress due to the transitory nature of seismic and wind loading. It is important to recognize that, in some cases, this one-third allowance has already been "built-in" to tables provided by the UBC and vendors. For example, the plywood diaphragm nailing requirements (as in Table 17) have already considered this increase, as have the connector/connection/strap/tie recommendations published by certain vendors. It is important to read the table footnotes to determine if this increase has already been included in published data.

88 COMBINED SEISMIC AND WIND LOADING

UBC provisions for ductile seismic detailing must always be met, even if the wind load is greater than the seismic load. Seismic loads generally control, but wind loads may control in seismic zone 3. It is not necessary for the building to be designed to withstand wind and seismic loads simultaneously [UBC Sec. 2330(b)].

Wind loading is covered by the UBC in Secs. 2311–2321. Specifically, the wind pressure on a structure is given by the following formula (corresponding to UBC Eq. 16-1).

$$P = C_e C_q q_s I \qquad [61]$$

C_e is a factor (from UBC Table 23-G) accounting for height, exposure, and gusts. C_q is a factor (from UBC Table 23-H) that depends on the type of structure exposed to the wind. q_s is the wind stagnation pressure (from UBC Table 23-F) at an elevation of 11 m (33 ft) (the standard value). q_s depends on wind speed. I is the importance factor for wind loading (from UBC Table 23-L).

The wind pressure can be assumed to act on the structure in one of two ways. The *normal force method* can be used for the design of any structure, including gabled

frames. Wind pressure is assumed to simultaneously act normal to all exterior surfaces [UBC Sec. 2317(b)]. The *projected area method* calculates the horizontal wind force by assuming the pressure acts upon the projected area of the structure. The projected area method can be used only if the building is less than 200 ft high and does not contain a gabled rigid frame [UBC Sec. 2317(c)].

Although codes such as AISC, AITC, and ACI provide for the combination of various loadings (e.g., snow and wind, earthquake and wind), seismic and wind loads are distinctly different in origin. Wind loads are applied over an exterior surface. Seismic loads are inertial in nature.

There is no such thing as a "governing" load when a building is in a potential earthquake area. In some cases, the maximum expected lateral wind loading will result in a larger drift (see Sec. 74) than an earthquake will. However, even in that instance, design must adhere to all provisions of the seismic code. The reason for this requirement is that the structure must resist seismic loads in a ductile manner even when it resists the wind load elastically.

89 WHEN THE UBC PERMITS A STATIC ANALYSIS

The UBC [Sec. 2333(h)] permits two methods to be used in determining the seismic loading: static and dynamic. (See Sec. 45.) The UBC is very specific about when the static method can be used. In general, any structure *may* be designed using the dynamic method at the option of the structural engineer, and some structures *must* use the dynamic method [Sec. 2333(h)3].

The static method may be used for buildings with the following characteristics:

⋄ Structures in seismic zones 1 and 2 (see Sec. 35) with standard occupancy category IV (see UBC Table 23-K), whether they are regular or irregular (see Sec. 101)[Sec. 2333(h)2A]

⋄ Regular structures under 73 m (240 ft) in height using one of the lateral force resisting systems listed in Table 14, Sec. 96 of this book (equivalent to UBC Table 23-O) except regular structures located on soil profile S_4 (see Table 13) which have natural periods greater than 0.7 s [Sec. 2333(h)2B]

⋄ Irregular structures less than or equal to five stories or 20 m (65 ft) in height [UBC Sec. 2333(h)2C]

⋄ Structures with flexible upper portions (e.g., towers) supported on a rigid lower portion if three conditions are met: (1) both portions, when considered individually, are regular, (2) the average story stiffness of the lower portion is at least ten times the average story stiffness of the upper portion, and (3) the period of the entire structure is no more than 1.1 times the period of the upper portion considered as a separate structure fixed at the base [UBC Sec. 2333(h)2D]

All structures not meeting these requirements, including irregular buildings (see Sec. 101), must be designed using the dynamic method [UBC Sec. 2333(h)3].

90 UBC BASE SHEAR CALCULATION

As described in Sec. 43, the *base shear*, V, is the total lateral inertial force imposed on the structure at its base by an earthquake. It is the sum of all the inertial story shears. Rather than calculating the story shears individually and then summing them to obtain the base shear, the UBC calculates the base shear from the total structure weight and then apportions the base shear to the stories. The base shear formula [UBC Sec. 2312(e)2A] is given by Eq. 62.

$$V = \frac{ZICW}{R_w} \qquad [62]$$

Equation 62 calculates the seismic force on the entire structure and cannot be used for parts (e.g., walls) of the structure. Equation 74 should be used for parts of the structure.

91 SEISMIC ZONE FACTOR: Z

The *seismic zone factor*, Z, accounts for the amount of seismic risk present in the building's seismic zone. There are six zones (see Sec. 35 and Fig. 11), with zone 0 representing the least risk. Table 11 (equivalent to UBC Table 23-I) gives the zone factors for each of the zones.[87]

Table 11
UBC Zone Factors (Z)
[UBC Table 23-I]

Zone	Z
0	0
1	0.075
2A	0.15
2B	0.20
3	0.30
4	0.40

The values of the zone factor are intended to represent the *effective* (not maximum) peak ground accelerations (see Sec. 23) that have only a 10% chance of being exceeded in 50 years (i.e., the *recurrence interval*). This corresponds to a ground motion that will be exceeded, on the average, only once in 475 years.

92 IMPORTANCE FACTOR: I

The *importance factor*, I, is either 1.0 or 1.25, depending on how critical it is for the structure to survive. Table 12 (equivalent to UBC Table 23-L) lists the values of the importance factor for different types of facilities.[88]

Table 12
UBC Importance Factor (I) for Earthquakes
[UBC Table 23-L]

Occupancy Category	Meaning	I
I	Essential facility	1.25
II	Hazardous facility	1.25
III	Special occupancy structure	1.00
IV	Standard occupancy structure	1.00

Essential facilities are emergency facilities that must remain operational after an earthquake [UBC Sec. 2331]. They include hospitals with surgery and emergency treatment facilities, fire and police stations, emergency preparedness structures (including structures housing emergency vehicles), and government communication centers required for emergency response. *Hazardous facilities* are used to store significant quantities of toxic and explosive substances that, if released, would be a health threat to the public. *Special occupancy structures* are designed to house or contain large quantities of people—for example, places of public assembly (300 or more persons), schools (250 or more students), colleges and adult education centers (500 or more students), jails, and hospitals without surgery or emergency capabilities. All other structures are considered to be *standard occupancy structures*. UBC Table 23-K defines the occupancy categories in greater detail.

---

[87]Development of the Z values was, at times, a compromise. For example, the two values for zone 2 represent the desire of building officials in the eastern United States to maintain historical force levels, regardless of the basis on which the zone boundaries were developed.

[88]The 1.25 maximum I value in the 1988 and later UBC is less than the 1.50 value in the 1985 UBC. This was a compromise which assumes that increased design and construction review, inspection, and ongoing observation will result in the same or higher levels of seismic performance in these structures. However, the requirements for increased amounts of review, inspection, and observation, while present in the Blue Book commentary, have been deleted in the UBC. This was not the original intent of the Seismology Committee of SEAOC.

93 C COEFFICIENT

The C coefficient (also referred to as the *base shear coefficient*) accounts for the period of vibration of the building and the supporting soil characteristics.[89] It is calculated from Eq. 63 (equivalent to UBC Formula 34-2).

$$C = \frac{1.25S}{T^{2/3}} \qquad [63]$$

The UBC [Sec. 2334(b)1] says that, except for code provisions requiring C to be scaled upward by a factor of 3, the minimum value of C/R_w is 0.075. The UBC [Sec. 2334(b)1] also says that C does not have to exceed 2.75 for any building, and that $C = 2.75$ can be used without regard to the soil type or building period. Figure 30 (equivalent to UBC Fig. 23-3) illustrates the variation of the C coefficient with building period, T, for various values of the site coefficient, S.

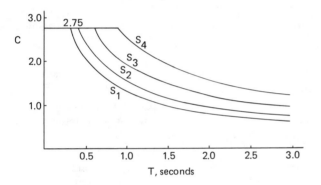

Figure 30 C Coefficient Versus Building Period and Soil Type

The value of C affects the base and story shears, which, in turn, determine the size of the lateral-force resisting elements. If method B (see Sec. 94) is used to find the period, T, the UBC [Sec. 2334(b)2B] requires that the value of C used for design cannot be less than 80% of the value that is determined from Eq. 63 using the empirical period Eq. 64 (i.e., method A). However, no such limit is placed on the value of C used to calculate the base and story shears, which, in turn, are used to calculate drift.

94 BUILDING PERIOD: T

The UBC gives two methods for determining the *building period*, T. The first is an approximate method.

---

[89]In the longer period ranges—greater than approximately 1.0 s—typical acceleration response spectra vary approximately as $1/T$. The two-thirds power in the denominator is a partial allowance for higher mode effects.

Equation 64 (corresponding to UBC Formula 34-3) can be used for all buildings.[90] In Eq. 64 (known as *method A*), h_n is the actual height (in feet) of the building above the base to the nth level. (The maximum allowable heights of the various types of structural systems are summarized in Table 14, Sec. 96 of this book, corresponding to UBC Table 23-O.)[91]

C_t is 0.035 for steel moment-resisting frames, 0.030 for reinforced concrete moment-resisting frames, 0.030 for eccentrically-braced frames (see Sec. 141), and 0.020 for all other buildings.[92]

$$T = C_t(h_n)^{3/4} \qquad [64]$$

The UBC [Sec. 2334(b)2A] contains an alternate method to be used in finding C_t for structures with concrete or masonry shear walls.[93]

Another method for finding the period, T, known as *method B*, is based on the deformation characteristics of the resisting elements and is a more rational determination. The UBC method B (UBC Eq. 34-5) is referred to by some engineers as the *Rayleigh method*. In addition to the UBC method of determining the period, any other substantiated analysis method can be used. If a dynamic analysis is performed, the first modal period should be used for T.

95 SITE COEFFICIENT: S

Table 13 (equivalent to UBC Table 23-J) lists the *site coefficient*, S, which is determined from the nature (known as the *soil profile*) of the soil. The site coefficient accounts for the ground motion amplification effect caused by softer soils. The soil profile must be determined by a properly substantiated geotechnical investigation. However, where soil properties are not known, soil profile S_3 must be used [UBC Table 23-J].

---

[90]This first method will probably be used for almost all preliminary designs and many final designs.

[91]The 1988 Blue Book recommended that masonry bearing wall systems in zones 3 and 4 be limited to 120 ft. The 1988 UBC set the limit at 160 ft. Thus, it is apparent that height limits are somewhat subjective. The discrepancy was eliminated in the 1990 Blue Book.

[92]Most of the data supporting these values came from the 1971 San Fernando earthquake.

[93]When Eq. 64 is used with shear wall structures, the base shear formula may be overly conservative in the lower seismic zones.

Profile S_4 is used only when a geotechnical study shows it to be valid or when required by the local building official.[94]

Table 13
UBC Site Coefficient (S)
[UBC Table 23-J]

Profile	Description	S
S_1	Either rock-like material with a shear-wave velocity in excess of 2500 ft/sec, or stiff or dense soil less than 200 ft thick	1.0
S_2	Stiff or dense soil greater than 200 ft thick	1.2
S_3	Soil 70 ft or more thick with 20 ft or more of soft to medium stiff clay, but not more than 40 ft of soft clay	1.5
S_4	Soil containing more than 40 ft of soft clay with a shear wave velocity less than 500 ft/sec	2.0

96 R_w FACTOR

The R_w factor accounts for the different energy-absorbing characteristics (including inelastic and damping effects) of the various types of structures in cyclic loading.[95] It is determined from the type of structural system used, as defined for buildings in Table 14 of this book (equivalent to UBC Table 23-O), and for nonbuildings in Table 16, Sec. 111 of this book (equivalent to UBC Table 23-Q). Systems with higher ductility (e.g., steel moment-resisting frames) have higher R_w values.

The UBC [Sec. 2333(f)] recognizes five different types of structural systems capable of resisting lateral forces. (These structural systems should not be confused with the *use* of the facility. Use is accounted for by the I factor.) (See Sec. 92.)

---

[94]The S_4 profile is appropriate in areas with large deposits of very soft clay that are subject to large amplifications in seismic ground motion. Buildings constructed on San Francisco Bay mud or on the Mexico City lake bed are likely candidates for the S_4 soil profile.

[95]The R_w factor is a partly empirical, partly judgmental factor that reduces the base shear to a predetermined value. This is the only "judgment factor" in the base shear equation.

A *bearing wall system* (called a *box system* in the 1985 UBC) is a system that relies on walls and/or braced frames to resist all of the loads on the structure. By itself, the word "wall" is ambiguous since there are two main types of structural walls. (Separation walls and curtains are not structural walls.) A *bearing wall* is a wall designed and constructed to resist vertical loads. A *shear wall* is a wall designed and constructed to resist lateral loads. A wall can also be used to resist both vertical and lateral loads. A bearing wall system does not have a complete vertical load-carrying frame. Bearing walls and braced frames support all of the gravity loads. Lateral forces are resisted by shear walls and/or braced frames. Bearing wall systems should not be confused with dual systems.

Bearing wall systems do not possess a complete vertical load-carrying frame and rely on walls or braced frames to carry both the vertical (gravity) and lateral (seismic) loads.[96] The distinguishing factor of these systems is that the failure of the primary seismic system also compromises the ability of the structure to support its dead and live loads.

A *frame (vertical load-carrying frame)* is a complete, self-contained, three-dimensional unit comprised of interconnected members. The frame carries vertical (gravity) loads. A frame may or may not resist lateral loads. If it does, it is known as a braced frame.

A *braced frame* is a truss system of interconnected members designed to resist lateral loads through the development of axial loads in the members. Braced frames can be of the concentric or eccentric types.

A *moment-resisting frame* carries both vertical and lateral loads. Lateral loads are resisted by deflection of the members and joints. Since joints are designed to be rigid, lateral building deflection is accompanied by flexure in the beams and columns. Moment-resisting frames can be constructed of either concrete or steel. There are three types of moment-resisting frames: *ordinary moment-resisting frames (OMRF)*, *intermediate moment-resisting frames (IMRF)*, and *special moment-resisting frame (SMRF)*.

Building frame systems use a complete space frame to carry the vertical (gravity) loads, but a separate system of nonbearing shear walls or braced frames resist

---

[96]It is still common to refer to this type of design as a *box system*, as it was described in earlier editions of the UBC.

the lateral (seismic) load.[97] Unlike the bearing wall systems, failure of the primary lateral support system does not compromise the ability of the structure to support its gravity loads.

Moment-resisting frames, [UBC Sec. 2333(f)4], resist forces in members and joints primarily by flexure and rely on a frame to carry both vertical and lateral loads. Lateral loads are carried primarily by flexure in the members and joints. Theoretically, joints are completely rigid. Special moment-resisting frames are specially detailed to ensure ductile behavior and comply with Chap. 26 (concrete) or Chap. 27 (steel) of the UBC. Intermediate moment-resisting frames, concrete frames with less stringent requirements designed in accordance with UBC Sec. 2625, cannot be used in seismic zones 3 and 4 [UBC Table 23-O, Ftn. 6]. Ordinary moment-resisting frames are steel or concrete moment-resisting frames that do not meet the special detailing requirements for ductile behavior. While ordinary moment-resisting frames constructed of steel can be used in any seismic zone, they cannot be constructed of concrete in seismic zones 2, 3, and 4 [UBC Table 23-O, Ftn. 7].

Dual systems (not to be confused with bearing wall systems) combine two of the previously mentioned systems to resist lateral loads. Specifically, a specially detailed moment-resisting frame of either concrete or steel serves in conjunction with shear walls or a braced frame. The moment-resisting space frame must be able to resist at least 25% of the base shear, and the two systems must be designed to resist the total lateral load in proportion to their relative rigidities [UBC Sec. 2333(f)5].[98] A complete frame resists all of the vertical loads.

Undefined structural systems do not fit into any of these categories. The designer of such systems must submit a rational basis for the design force level used [UBC Sec. 2333(f)6].

Nonbuilding structures are defined in the UBC [Sec. 2333(f)7]. (See Sec. 111.)

---

[97]A *space frame* is a three-dimensional system consisting of interconnected members that operate as a single unit. A *frame* is a truss-like two-dimensional system with concentric or eccentric connection points in which the lateral forces are resisted by axial stresses in the members. The term *space* is generally not appropriate because lateral loads are resisted by two-dimensional structures. The term *space*, although present in the 1988 Blue Book and UBC, was removed from the terminology in the 1990 Blue Book and 1991 UBC.

[98]Remember that the moment-resisting frame can be either steel or concrete, but intermediate frames cannot be constructed of concrete in seismic zones 3 and 4 [UBC Table 23-O, Ftn. 6].

97 WEIGHT: W

The weight, W (in pounds), used in Eq. 62 is normally the total dead load of the structure.[99] This includes the weight of the ceiling, partitions, pipes, ducts, and equipment that is normally attached. However, there are cases in which other loads must be included as well. These cases and the loads are described below.

⬦ A minimum of 25% of the floor live load (i.e., storage) is added in warehouses and storage buildings [UBC Sec. 2334(a)1].

⬦ Not less than 10 pounds per square foot must be added when partition loads are used in the design of the floor [UBC Sec. 2334(a)2].

⬦ Snow loads exceeding 30 pounds per square foot must be included, unless a reduction of up to 75% is approved by the local building official [UBC Sec. 2334(a)3].

⬦ The total weight of permanent equipment must be included [UBC Sec. 2334(a)4].

Strictly speaking, the units of building weight, W, will determine the units of base shear, V. Thus, if weight is expressed in kips, the base shear will be in kips.

Mass is seldom, if ever, used in practice, as all weights are given in pounds or kips. However, care must be observed in the unlikely event that the building *mass*, as opposed to the building *weight*, is specified. Equation 65 shows that the weight in pounds-force (lbf) is numerically the same as the mass in pounds-mass (lbm), although this is not true if other units of mass are used.

$$W_{\text{pounds-force}} = m_{\text{pounds-mass}} \left(\frac{g}{g_c} \right) \qquad [65]$$

$$W_{\text{pounds-force}} = m_{\text{slugs}} g \qquad [66]$$

$$g = 32.2 \, \frac{\text{ft}}{\text{sec}^2}$$

$$g_c = 32.2 \, \frac{\text{ft-lbm}}{\text{lbf-sec}^2}$$

$$W_{\text{newtons}} = m_{\text{kilograms}} g \qquad [67]$$

$$g = 9.81 \, \frac{\text{m}}{\text{s}^2}$$

---

[99]As part of normal practice, all of the at-grade weight and half of the first-story wall weight are commonly omitted in the analysis of seismic diaphragm loads. (See Sec. 116.) However, such a provision is not explicitly defined in the UBC for calculating base shear, nor is such a provision mentioned in the Blue Book commentary.

Table 14
Structural Systems and R_w Values
[UBC Table 23-O]

BASIC STRUCTURAL SYSTEM[1]	LATERAL LOAD-RESISTING SYSTEM—DESCRIPTION	R_w[2]	H[3]
A. Bearing Wall System	1. Light-framed walls with shear panels		
	a. Plywood walls for structures three stories or less	8	65
	b. All other light-framed walls	6	65
	2. Shear walls		
	a. Concrete	6	160
	b. Masonry	6	160
	3. Light steel-framed bearing walls with tension-only bracing	4	65
	4. Braced frames where bracing carries gravity loads		
	a. Steel	6	160
	b. Concrete[4]	4	—
	c. Heavy timber	4	65
B. Building Frame System	1. Steel eccentrically braced frame (EBF)	10	240
	2. Light-framed walls with shear panels		
	a. Plywood walls for structures three stories or less	9	65
	b. All other light-framed walls	7	65
	3. Shear walls		
	a. Concrete	8	240
	b. Masonry	8	160
	4. Concentrically braced frames		
	a. Steel	8	160
	b. Concrete[4]	8	—
	c. Heavy timber	8	65
C. Moment-resisting Frame System	1. Special moment-resisting frames (SMRF)		
	a. Steel	12	N.L.
	b. Concrete	12	N.L.
	2. Concrete intermediate moment-resisting frames (IMRF)[6]	8	—
	3. Ordinary moment-resisting frames (OMRF)		
	a. Steel	6	160
	b. Concrete[7]	5	—
D. Dual Systems	1. Shear walls		
	a. Concrete with SMRF	12	N.L.
	b. Concrete with steel OMRF	6	160
	c. Concrete with concrete IMRF[6]	9	160
	d. Masonry with SMRF	8	160
	e. Masonry with steel OMRF	6	160
	f. Masonry with concrete IMRF[4]	7	—
	2. Steel EBF		
	a. With steel SMRF	12	N.L.
	b. With steel OMRF	6	160
	3. Concentrically braced frames		
	a. Steel with steel SMRF	10	N.L.
	b. Steel with steel OMRF	6	160
	c. Concrete with concrete SMRF[4]	9	—
	d. Concrete with concrete IMRF[4]	6	—
E. Undefined Systems	See Sections 2333 (h) 3 and 2333 (i) 2	—	—

[1]Basic structural systems are defined in Section 2333 (f).
[2]See Section 2334 (c) for combination of structural system.
[3]H—Height limit applicable to Seismic Zones Nos. 3 and 4. See Section 2333 (g).
[4]Prohibited in Seismic Zones Nos. 3 and 4.
[5]N.L.—No limit.
[6]Prohibited in Seismic Zones Nos. 3 and 4, except as permitted in Section 2338 (b).
[7]Prohibited in Seismic Zones Nos. 2, 3 and 4.

Example 10

A 100-ft-tall, 10-story office building has a total weight of 15,000 kips. The building is located in seismic zone 4 and is constructed on soil type 2. It is designed with a special moment-resisting steel frame system. Use the UBC to calculate the design base shear.

Solution

From Table 11 for zone 4, the zone factor, Z, is 0.40. From Table 12, the occupancy importance factor, I, is 1.00. From Table 13, the site coefficient, S, is 1.2. From Table 14 for a steel moment-resisting frame, $R_w = 12$. For a steel moment-resisting frame, $C_t = 0.035$.

Equation 63 gives the natural period as a function of the building height.

$$T = C_t(h_n)^{\frac{3}{4}} = (0.035)(100\,\text{ft})^{\frac{3}{4}} = 1.107\,\text{sec}$$

From Eq. 63, the C coefficient is

$$C = \frac{1.25S}{T^{\frac{2}{3}}} = \frac{(1.25)(1.2)}{(1.107\,\text{sec})^{\frac{2}{3}}} = 1.402$$

Equation 62 gives the base shear.

$$V = \frac{ZICW}{R_w} = \frac{(0.40)(1.00)(1.402)(15,000\,\text{kips})}{12}$$
$$= 701\,\text{kips}$$

98 SIGNIFICANCE OF ZC

The product, ZC, of UBC terms Z and C as a function of building period, T, is essentially a multimode acceleration response spectrum (see Sec. 65) for major ground motion. This is illustrated in Fig. 31, which also illustrates the reduction in design spectrum caused by the R_w coefficient.

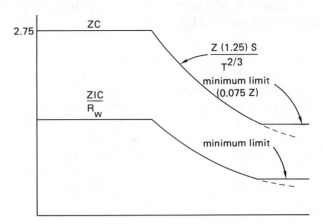

Figure 31 ZC Response Spectrum

The maximum value of ZC in seismic zone 4 is $(0.4)(2.75) = 1.1$. This represents a peak ground acceleration of 1.1 g.

99 UBC PROVISIONS FOR DIRECTION OF THE EARTHQUAKE

The UBC requires in Sec. 2334(a) that the lateral forces from any horizontal direction (i.e., x- and y-direction) be evaluated separately for regular structures. Usually directions parallel and perpendicular to the major building faces, regardless of the building's orientation with respect to a major fault, will be used. It is assumed that a building will have enough lateral strength to resist an *oblique earthquake* when requirements for these two orthogonal directions have been met.

A simultaneous application of earthquake forces from two directions could overstress parts of the lateral force-resisting system in certain cases, such as when there is torsional irregularity or a nonparallel structural system, or when a given member—usually a corner column—is part of two intersecting lateral force-resisting systems. UBC Sec. 2337(a) requires an analysis of such directional effects.

The UBC seems to require an infinite number of analyses by requiring the structure to be designed for forces coming from "...any horizontal direction" [UBC Sec. 2334(a)]. However, the next sentence clarifies this by specifying that, except for certain types of irregular buildings, the "... seismic forces may be assumed to act noncurrently [sic] in the direction of each principal axis..." The UBC does not generally require the structure to be designed for earthquakes coming from more than these two orthogonal directions.

100 UBC PROVISIONS FOR VERTICAL GROUND ACCELERATION

The UBC contains provisions in Sec. 2334(j) for determining the effects of vertical acceleration on cantilever members and long-span prestressed elements only in seismic zones 3 and 4. Specifically, horizontal cantilevers must be designed for a net upward force of 20% of their own weight. Only 50% of the dead weight of horizontal prestressed members can be used or considered in their design (i.e., the dead load may be removed, driving the upper surface of horizontal beams into the tension region).

101 IRREGULAR BUILDINGS

Buildings with irregular shapes, changes in mass from floor to floor, variable stiffness with height, and unusual setbacks do not perform well during earthquakes. This is unfortunate from aesthetic and creative perspectives, because such buildings are generally among the most pleasing in appearance and interesting to design. The UBC [Sec. 2333(e)1] requires that all buildings be classified as *regular* or *irregular*. With few exceptions (see Sec. 89), the code specifies that a dynamic analysis is required for an irregular building.[100]

It is significant that if a static analysis shows that the story-drifts are substantially linear, then some buildings can be recategorized as vertically regular [UBC Sec. 2333(e)3B]. Thus, it is the drift that determines vertical irregularity, not the plan view.

102 VERTICAL STRUCTURAL IRREGULARITIES

UBC Table 23-M lists and defines the following five types of *vertical structural irregularities.*

1. A *soft story* [Sec. 2333(e)3] has a stiffness less than 70% of the story immediately above, or less that 80% of the average stiffness of the three stories above.

2. A story has a *mass (weight) irregularity* [Sec. 2333(e)3] when its mass is more than 150% of the mass of a story above or below. (Roofs lighter than the floor immediately below are excluded.)

3. A story has *vertical geometric irregularity* [Sec. 2333(e)3] when the horizontal dimension of a story's lateral force-resisting system is more than 130% of that in an adjacent story. (One-story penthouses are excluded.)

4. An *in-plane discontinuity* [Sec. 2334(g)] exists at a story when there is an in-plane offset of the lateral load-resisting elements greater than the length of those elements.

5. A *weak story* [Sec. 2334(i)1] exists when the story strength is less than 80% of that in the story above. The *story strength* is defined as the strength of all of the seismic resisting elements sharing the story shear in the direction of the earthquake.

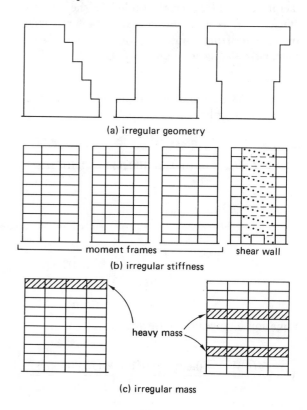

Figure 32   Vertical Irregularities

A structure that meets one of the five conditions of UBC Table 23-M (corresponding to the five conditions listed in Sec. 102) would normally be considered irregular. However, a structure that would otherwise be considered irregular under the first two conditions listed in Sec. 102 can be considered regular if the story drift ratio (as calculated from the design lateral forces and neglecting torsional effects) for each floor is less than 1.3 times the story drift ratio for the floor above. (The story drift ratio is the ratio of the actual drift relative to

---

[100]While the code is specific, some of the decision criteria were set primarily by judgment, as neither historical or empirical data were available. Thus, the criteria are somewhat "loose," and the intent is to affect only the worst cases of structural irregularity and to allow less severe structural configurations.

the floor below and the floor-to-floor height.) Furthermore, this condition does not even have to be satisfied by the top two stories as long as all stories below the top two satisfy it [UBC Sec. 2333(e)3B].

103 PLAN STRUCTURAL IRREGULARITIES

UBC Table 23-N lists and defines the following five types of *plan structural irregularities*.

1. *Torsional irregularity* [Sec. 2337(b)9E] exists when the maximum story drift (caused by the lateral load *and* the accidental torsion) at one end of the structure transverse to its axis is more than 1.2 times the average story drifts calculated from both ends. Only buildings with rigid diaphragms (Sec. 118) are affected by this type of irregularity.[101]

2. A building has *reentrant corner irregularity* [Secs. 2337(b)9E and 2337(b)9F] when one or more parts of the structure project beyond a reentrant corner a distance greater than 15% of the plan dimension in the given direction.

3. *Diaphragm discontinuity* [Sec. 2337(b)9E] occurs with diaphragms having abrupt discontinuities or variations in stiffness, including when there are cutout, or open, areas greater than 50% of the gross diaphragm area, or when the stiffness of the diaphragm changes more than 50% from story to adjacent story.

4. An *out-of-plane offset* [Secs. 2334(g) and 2337(b)9E] is a discontinuity in the lateral force path—an out-of-plane offset of the vertical elements.

5. A *non-parallel system* [Sec. 2337(a)] is one for which the vertical load-carrying elements are not parallel to or symmetrical about the major orthogonal axes of the lateral force-resisting system. This also includes buildings in which a column is part of two or more intersecting lateral force-resisting systems (as would most likely occur in a corner column at the ends of two orthogonal frames), unless the axial column load due to seismic forces is less than 20% of the column allowable load.

---

[101] As defined in UBC Sec. 2334(f), a *flexible diaphragm* is one that has a maximum lateral deflection at a story more than two times the average story drift at that story.

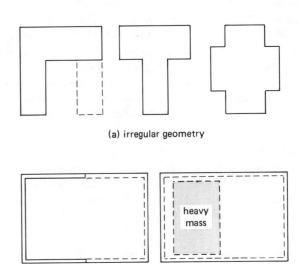

(a) irregular geometry

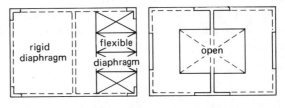

(b) irregular mass/resistance

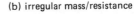

(c) irregular diaphragm stiffness

Figure 33   Plan Irregularities

104 PENALTIES FOR STRUCTURAL IRREGULARITY

The so-called "penalty" for some irregular buildings is the requirement of a dynamic analysis. For example, buildings greater than five stories in height with vertical mass, geometry, or stiffness irregularities (types A, B, C in UBC Table 23-M) must receive a dynamic treatment.

However, the other types of irregularity are not adequately mitigated merely by performing a dynamic analysis. The UBC penalizes irregular structures of these types both by imposing additional requirements and by eliminating special allowances permitted regular structures. For example, the total base shear can be reduced by 10% for regular structures when a dynamic analysis is performed according to the method specified by the UBC [Sec. 2335(e)3A(ii)]. When a structure is irregular, this bonus is not permitted.

As additional examples of how irregularity is discouraged, *weak stories* are prohibited [UBC Sec. 2333(i)1] in buildings over two stories in height when the strength ratio based on the story above is below 65%. Heights are more limited for irregular structures [UBC Sec. 2334(i)]. Discontinuous lateral force-resisting systems (i.e., discontinuous shear walls) must receive special checking and detailing to ensure ductile behavior [UBC Sec. 2334(g)].

The omission of the higher mode F_t force in overturning moment calculations is not permitted for irregular structures in calculations of soil pressure and foundation design [UBC Sec. 2910(d)].

Torsional irregularity is penalized by requiring the use of an increased accidental eccentricity [UBC Sec. 2334(f)]. Reduced stress limits are required for connections of diaphragms to vertical elements and to drag members and between drag members themselves with plan irregularities [UBC Sec. 2337(b)9E]. Seismic forces in two orthogonal directions must be evaluated when torsional irregularities exist in both principal horizontal directions [UBC Sec. 2337(a)].

105 DISTRIBUTION OF BASE SHEAR TO STORIES

The base shear, V, is distributed to the n stories in accordance with Eqs. 68 and 70 (corresponding to UBC Formulas 34-7 and 34-8). The F_x forces increase linearly with height above the base, as Fig. 34 illustrates.[102] F_t is an additional force that is applied to the top level (i.e., the roof) in addition to the F_x force at that level. The F_t accounts for higher-mode effects. It is zero for $T \leq 0.7$ s.

$$F_t = 0.07TV \quad [F_t < 0.25V] \qquad [68]$$

$$F_t = 0 \quad [T \leq 0.7 \text{ s}] \qquad [69]$$

$$F_x = \frac{(V - F_t)W_x h_x}{\sum\limits_{i=1}^{n} W_i h_i} \qquad [70]$$

$$V = F_t + \Sigma F_x \qquad [71]$$

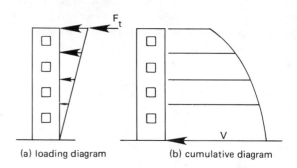

Figure 34   Distribution of Story Shears

As Fig. 34 shows, the distribution of the F_x is linear.[103] This distribution depends on the following assumptions. Buildings that do not meet these requirements must receive a dynamic analysis.

1. The building is regular and nearly symmetrical.

2. The lateral stiffnesses of each floor are approximately the same.

3. The lateral stiffnesses of each floor are uniformly distributed in plan.

4. The weight of each floor is approximately the same and is uniformly distributed in plan.

5. There is a smooth force path between members.

6. Torsional components of drift are small compared to translational components.

7. The first deflection mode (see Sec. 60) can be approximately represented as a straight line.

◇   ◇   ◇   ◇   ◇   ◇   ◇

Example 11

A 5-story building is constructed with 12-ft story heights. The base shear has been calculated as 160 kips. Each story floor has a weight of 800 kips, and the roof has a weight of 700 kips. The natural period of oscillation is 0.5 sec. (a) What are the story shears? (b) What are the effective spectral accelerations (S_a) in gravities at each floor?

Solution

The top force, F_t, is zero since $T < 0.7$ sec.

---

[102]This is sometimes referred to as "throwing weight to the top."

[103]There have been some attempts, such as in the ATC and NEHRB documents, to introduce an exponent that varies from linear to quadratic as a function of the building period. There is no consensus that this improves the results, although it certainly increases the complexity.

A table is the easiest way to set up the data for calculating and recording the story shears and accelerations.

level x	h_x	W_x	$h_x W_x$	$\dfrac{h_x W_x}{\sum h_x W_x}$	F_x (kips)	S_a (g's)
5 (roof)	60	700	42,000	0.304	48.6	0.069
4	48	800	38,400	0.278	44.5	0.056
3	36	800	28,800	0.209	33.4	0.042
2	24	800	19,200	0.139	22.2	0.028
1	12	800	9,600	0.070	11.2	0.014
TOTALS		3900	138,000	1.000	160.0	

The value of F_5 is given by Eq. 70.

$$F_5 = (V - F_t)\frac{W_5 h_5}{\Sigma W_i h_i}$$
$$= (160 \text{ kips} - 0 \text{ kips})(0.304)$$
$$= 48.6 \text{ kips}$$

The effective spectral acceleration is the actual acceleration experienced by the floor. (See Sec. 42.) It can be calculated from Newton's second law and the force and weight. (See Sec. 43.) For the roof, there are two forces acting.

$$S_a = \frac{F}{W} \quad [S_a \text{ in gravities}]$$
$$S_{a,5} = \frac{F_t + F_5}{W_5} = \frac{0 \text{ kips} + 48.6 \text{ kips}}{700 \text{ kips}}$$
$$= 0.069 \text{ gravities}$$

Calculations for lower floors are similar but do not include the top force, F_t, in the calculation of spectral acceleration.

106 UBC PROVISIONS FOR TORSION

The UBC requires in Sec. 2334(e) that a certain amount of *accidental torsion* (see Sec. 76) be planned for, even in regular buildings. Specifically, the *center of mass* at each level is assumed to be displaced from the calculated center of mass a distance of 5% of the building direction perpendicular to the direction of the seismic force. Thus, the *accidental eccentricity* will be different for the two orthogonal directions.

The accidental torsion is not a minimum value as it was in previous UBCs. It is a value to be added to the actual calculated eccentricity [UBC Sec. 2334(f)].

The accidental torsion may have to be increased above the 5% level for torsionally-irregular buildings by use of an amplification factor [UBC Sec. 2334(f)].

The term *design eccentricity* is used to represent the sum of the actual and accidental (5%) eccentricities.

107 UBC PROVISIONS FOR MAXIMUM DRIFT

Unless it can be shown that greater amounts can be tolerated, *story drift* (the drift of one story relative to the story below) in buildings with periods less than 0.7 s is limited to $0.04/R_w$ or 0.005 times the story height (i.e., floor-to-ceiling height), whichever is less. Story drift in buildings with periods greater than 0.7 s is limited to $0.03/R_w$ or 0.004 times the story height, whichever is less [UBC Sec. 2334(h)2].

It should be noted that the UBC drift provisions are not absolute. If greater drifts can be tolerated, they are acceptable [UBC Sec. 2334(h)2].[104]

108 UBC PROVISIONS FOR OVERTURNING MOMENT

The UBC requires that, at any level, the *overturning moment* (see Sec. 133) must be determined using the seismic forces that act on all of the levels above [UBC Sec. 2334(g)1].[105] These forces are F_t and F_x as defined in Eqs. 68 and 70. The effect of *uplift* must also be checked. Any net tension must be resisted by interaction with the soil (e.g., friction piles that resist uplift).

In regular structures, the top force, F_t, can be omitted in calculating the overturning effects at the soil-foundation interface, including the calculation of soil pressure under spread footings and the soil-pile frictional forces during uplift [UBC Sec. 2910(d)].[106] This

---

[104]For example, many prefabricated, single-story steel buildings (commonly referred to as "Butler buildings") do not meet the code drift limit. Such deviations from the code may be permitted, particularly in industrial or warehouse space.

[105]There are those who believe it is unnecessary to use 100% of all the story shears (i.e., the F_x values) in calculating the overturning moment because the maximum forces actually occur at different times, since they correspond to the modal shapes. Thus, reductions up to 50% might actually be appropriate. However, the Blue Book and UBC do not include any such reduction for structural calculations when a static analysis is performed except when calculating the overturning moment at the soil-foundation interface in regular structures [UBC Sec. 2910(d)]. The only way to "get" the reduction is to perform a dynamic analysis, even when it is not required.

[106]The omission of F_t is not permitted, however, in the design of structural or foundation elements, including the design of the footings and piles that resist uplift and of their connections with the building.

omission is permitted because the F_t force represents higher mode forces, and the higher modes, regardless of the modal phasing, do not contribute to overturning at the base.

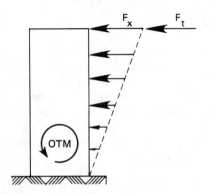

Figure 35   Overturning Moment at the Base

109 UBC PROVISIONS FOR P-Δ EFFECT

The UBC requires [Sec. 2334(i)] that the forces and story drifts be used to calculate the overall structural frame stability due to the P-Δ effect unless the structure is in seismic zones 3 or 4 and the design satisfies the drift limitations (Sec. 107). The Blue Book commentary introduces the concept of a *stability coefficient*, θ, which is the ratio of secondary to primary moments. If this stability coefficient is less than 0.10, the P-Δ effect can be disregarded. Otherwise, a rational analysis must be used to evaluate P-Δ.

$$\theta = \frac{P_x \Delta}{V_x h_{sx}} \qquad [73]$$

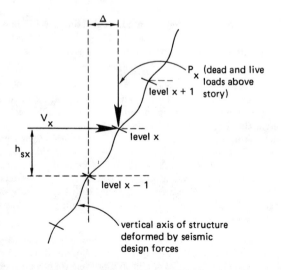

Figure 36   Stability Coefficient Variables

110 SEISMIC FORCE ON PARTS AND PORTIONS

This section covers the calculation of the force on a rigid part of a building. This should not be confused with the calculation of the force on an entire rigid nonbuilding. (See Sec. 112.)

The seismic loading, F_p, on a *part* or *portion* is calculated from Eq. 73 (corresponding to UBC Formula 36-1). (Parts or portions are such elements as signs, unbraced parapet walls, interior non-bearing walls and partitions, roof-mounted air conditioning equipment, water tanks, and penthouses.) This equation is used to design the vertical elements, their connections and anchorages, and members and connections that transfer these forces into the building's lateral force-resisting system. Values of Z and I are generally the same as those used for the building, except that $I = 1.5$ for machinery and equipment required for life-safety systems, as well as for tanks containing toxic and explosive substances [UBC Sec. 2336(b)]. W_p is the weight of the part or portion. The soil conditions and the natural period of the part are not factors in calculating the force on the part.

$$F_p = ZIC_pW_p \qquad [73]$$

Values of the coefficient C_p are given in Table 15 (corresponding to UBC Table 23-P) for rigid elements. *Rigid elements* and *rigid equipment* (not to be confused with rigid structures discussed in Sec. 111) are defined as those having natural periods of oscillation about their own fixed bases of 0.06 s or less. This includes exterior wall panels and building cladding. *Nonrigid elements* (*flexibly supported elements*) have periods in excess of 0.06 s. The value of C_p in Table 15 can simply be doubled for nonrigid elements, or a detailed (i.e., dynamic) analysis can be performed. However, the factor for nonrigid elements evaluated by dynamic analysis must be at least the value in Table 15. In no case for rigid or nonrigid elements does the value of C_p have to exceed 2.0 [UBC Sec. 2336(b)].

The seismic force on a diaphragm is not calculated in the same manner as any other part or portion. That is, $F_p = ZIC_pW_p$ is not used. Rather, the UBC Eq. 37-1 [Sec. 2337(b)9B] specifies the method to calculate the force on the part (i.e., diaphragm) at level x.

$$F_{px} = \left(\frac{F_t + \sum\limits_{i=x}^{n} F_i}{\sum\limits_{i=x}^{n} W_i} \right) (W_{px}) \qquad [74]$$

Table 15
Values of C_p for Rigid Parts and Portions[1]
[UBC Table 23-P]

ELEMENTS OF STRUCTURES AND NONSTRUCTURAL COMPONENTS AND EQUIPMENT[1]	VALUE OF C_p	FOOTNOTE
I. **Part or Portion of Structure**		
1. Walls including the following:		
a. Unbraced (cantilevered) parapets	2.00	
b. Other exterior walls above the ground floor	0.75	2,3
c. All interior bearing and nonbearing walls and partitions	0.75	3
d. Masonry or concrete fences over 6 feet high	0.75	
2. Penthouse (except when framed by an extension of the structural frame)	0.75	
3. Connections for prefabricated structural elements other than walls, with force applied at center of gravity	0.75	4
4. Diaphragms	—	5
II. **Nonstructural Components**		
1. Exterior and interior ornamentations and appendages	2.00	
2. Chimneys, stacks, trussed towers and tanks on legs:		
a. Supported on or projecting as an unbraced cantilever above the roof more than one half their total height	2.00	
b. All others, including those supported below the roof with unbraced projection above the roof less than one half its height, or braced or guyed to the structural frame at or above their centers of mass	0.75	
3. Signs and billboards	2.00	
4. Storage racks (include contents)	0.75	10
5. Anchorage for permanent floor-supported cabinets and book stacks more than 5 feet in height (include contents)	0.75	
6. Anchorage for suspended ceilings and light fixtures	0.75	4,6,7
7. Access floor systems	0.75	4,9
III. **Equipment**		
1. Tanks and vessels (include contents), including support systems and anchorage	0.75	
2. Electrical, mechanical and plumbing equipment and associated conduit, ductwork and piping, and machinery	0.75	8

[1]See Section 2336 (b) for items supported at or below grade.

[2]See Section 2337 (b) 4 C and Section 2336 (b).

[3]Where flexible diaphragms, as defined in Section 2334 (f), provide lateral support for walls and partitions, the value of C_p for anchorage shall be increased 50 percent for the center one half of the diaphragm span.

[4]Applies to Seismic Zones Nos. 2, 3 and 4 only.

[5]See Section 2337 (b) 9.

[6]Ceiling weight shall include all light fixtures and other equipment or partitions which are laterally supported by the ceiling. For purposes of determining the seismic force, a ceiling weight of not less than four pounds per square foot shall be used.

[7]Ceilings constructed of lath and plaster or gypsum board screw or nail attached to suspended members that support a ceiling at one level extending from wall to wall need not be analyzed provided the walls are not over 50 feet apart.

[8]Machinery and equipment include, but are not limited to, boilers, chillers, heat exchangers, pumps, air-handling units, cooling towers, control panels, motors, switch gear, transformers and life-safety equipment. It shall include major conduit, ducting and piping serving such machinery and equipment and fire sprinkler systems. See Section 2336 (b) for additional requirements for determining C_p for nonrigid or flexibly mounted equipment.

[9]W_p for access floor systems shall be the dead load of the access floor system plus 25 percent of the floor live load plus a 10 psf partition load allowance.

[10]In lieu of the tabulated values, steel storage racks may be designed in accordance with U.B.C. Standard No. 27-11.

The diaphragm should be designed to resist force F_p. However, regardless of the value calculated from UBC Eq. 37-1, F_p may not be less than $0.35ZIW_{\text{diaphragm}}$, nor need the force exceed $0.75ZIW_{\text{diaphragm}}$ ($W_{\text{diaphragm}}$ is the weight of the diaphragm plus any elements tributary to it) [Sec. 2337(b)9B]. Thus, $0.35ZIW_{\text{diaphragm}}$ should be taken as the minimum diaphragm force at any level.

Exterior walls that do not carry vertical or shear loads must resist a seismic force F_p calculated from Eq. 74. Connectors (e.g., bolts, inserts, welds, and dowels) must be sized for $4F_p$, while connection bodies (e.g., angles, bars, and plates) must be designed for $(4/3)F_p$. Drift, movement, and other standards also apply [UBC Sec. 2337(b)4B].

The 1991 UBC includes a new provision (Ftn. 3 in UBC Table 23-P) for walls whose lateral support comes from flexible diaphragms. The value of C_p used to calculate seismic force on a wall must be increased 50% (from 0.75 to 1.125) in the center one-half of the diaphragm span. This will be relevant when detailing the wall-to-diaphragm connectors.

This requirement stems from new data derived from observations and instrumentation of diaphragms in masonry buildings during pre-1992 earthquakes. The evidence seems to indicate that deflections at the midspans of large wood diaphragms may be two to three times larger than at the ends (where the parallel walls provide support). In the 1971 San Fernando earthquake, there was separation at the parallel panel joints as wide as two panel widths from the walls. This could have been prevented by tying across the panel joints, but ties are not required by the UBC at such a distance from the wall.

Example 12

A parapet wall extends 4 ft above the roof line of a one-story concrete-walled commercial building, as shown. The walls are 8 in thick. Consider all connections between the roof, walls, and footings to be pinned (free to rotate). The building is located in seismic zone 4. An earthquake acts perpendicular to the wall face. Determine (a) the force that the wall-roof connection must be designed to withstand and (b) the moment at the base of the parapet wall.

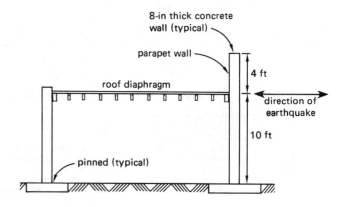

Solution

(a) The wall width is not given, so work with a 1-ft strip of wall. Concrete weighs approximately 150 lbf/ft^3, so the distributed weight per foot of width is

$$
\begin{aligned}
W_p &= \gamma \times \text{volume} \\
&= \gamma \times \text{width} \times \text{height} \times \text{thickness} \\
&= \frac{\left(150\,\dfrac{\text{lbf}}{\text{ft}^3}\right)(1\,\text{ft})(14\,\text{ft})(8\,\text{in})}{12\,\dfrac{\text{in}}{\text{ft}}} \\
&= 1400\,\text{lbf [per foot of wall]}
\end{aligned}
$$

While only the force on the parapet is wanted, the entire wall must first be analyzed. From Table 11, the seismic zone coefficient, Z, is 0.40. From Table 12, the occupancy importance coefficient is 1.00. From Table 15, $C_p = 0.75$. From Eq. 73, the distributed seismic force per foot of wall width is

$$
\begin{aligned}
F_p &= ZIC_pW_p = (0.40)(1.00)(0.75)(1400\,\text{lbf}) \\
&= 420\,\text{lbf [per foot of wall]}
\end{aligned}
$$

For the purpose of finding reactions, this force can be assumed to act at the mid-height of the wall, at $14/2$ ft = 7 ft. Summing moments about the wall base,

$$
\Sigma M = (7\,\text{ft})\left(420\,\frac{\text{lbf}}{\text{ft}}\right) - (F_{\text{anchor}})(10\,\text{ft}) = 0
$$

$$
F_{\text{anchor}} = 294\,\text{lbf [per foot of wall]}
$$

Since the UBC [Sec. 2337(b)8] requires the roof-wall diaphragm connection to withstand a minimum force of 200 lbf/ft [Sec. 2310], the calculated value of 294 lbf/ft controls. (See Sec. 149.)

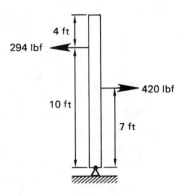

(b) The weight of the parapet alone is

$$W_p = \gamma \times \text{volume}$$
$$= \gamma \times \text{width} \times \text{height} \times \text{thickness}$$
$$= \frac{\left(150\,\dfrac{\text{lbf}}{\text{ft}^3}\right)(1\,\text{ft})(4\,\text{ft})(8\,\text{in})}{12\,\dfrac{\text{in}}{\text{ft}}}$$
$$= 400\,\text{lbf [per foot of wall]}$$

From Table 15 for the parapet alone, $C_p = 2.00$. From Eq. 73, the distributed seismic force per foot of wall width is

$$F_p = ZIC_pW_p = (0.40)(1.00)(2.00)(400\,\text{lbf})$$
$$= 320\,\text{lbf [per foot of wall]}$$

For the purpose of determining the moment at the base of the parapet, this force can be assumed to act at the mid-height, 2 ft up from the base. In effect, the parapet acts as a vertical cantilever wall. The net moment at the parapet base (i.e., where it joins the roof) is

$$M = (2\,\text{ft})(320\,\text{lbf}) = 640\,\text{ft-lbf [per foot of wall]}$$

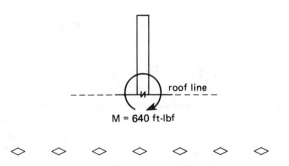

111 NONBUILDING STRUCTURES

Nonbuilding structures [UBC Sec. 2338] are significant self-supporting structures that are not housed within buildings but, nevertheless, come within the jurisdiction of the local building official.[107] Covered by the UBC are structures that (1) have building-like structural systems such as those described in UBC Sec. 2333(f), (2) are rigid systems (i.e., with periods less than 0.06 s), or (3) are specifically mentioned in UBC Table 23-Q. (To qualify as having a *building-like structural system*, the structure must have one or more levels (floors and roof) at which the mass is concentrated, and the framing system must extend between the levels.)

Most nonbuilding structures, even though they are not designed to accommodate people, are supported by structural systems traditionally found in occupied buildings. That is the reason the lateral forces on nonbuilding structures are calculated in the same manner as those for building structures. Rigid nonbuilding structures, however, are handled differently.

Rigid nonbuilding structures (not to be confused with rigid parts and portions of normal buildings) are structures with natural periods of less than 0.06 s. The natural period is the determining factor of whether the structure is rigid. An example would be a concrete pedestal structure.

Rigid structures are covered in UBC Sec. 2338(b). The lateral force, V, on rigid structures and their anchorages is given by UBC Formula 38-1, which assigns the value of $C/R_w = 0.5$.

$$V = 0.5ZIW \qquad [75]$$

The lateral force is distributed according to the distribution of the mass. It is assumed to act in any direction.

The basic base shear formula, Eq. 76, is used for nonbuilding structures with periods greater than 0.06 s. The weight, W, includes the weight of the full contents (if any) of the structure [UBC Sec. 2338(a)3]. Values of R_w are given in Table 16 (corresponding to UBC Table 23-Q).[108]

$$V = \frac{ZICW}{R_w} \qquad [76]$$

$$C = \frac{1.25S}{T^{2/3}} \qquad [77]$$

---

[107]Other items specifically not included in the UBC are offshore platforms, electrical transmission towers, dams, and highway and railroad bridges. These structures are not normally within the jurisdiction of the local building official.

[108]The values of R_w for nonbuilding structures are generally less than for buildings. This is considered justified because nonbuildings do not have the structural redundancy of multiple bays and nonstructural panels that effectively give buildings greater strength and damping than is considered in the design process.

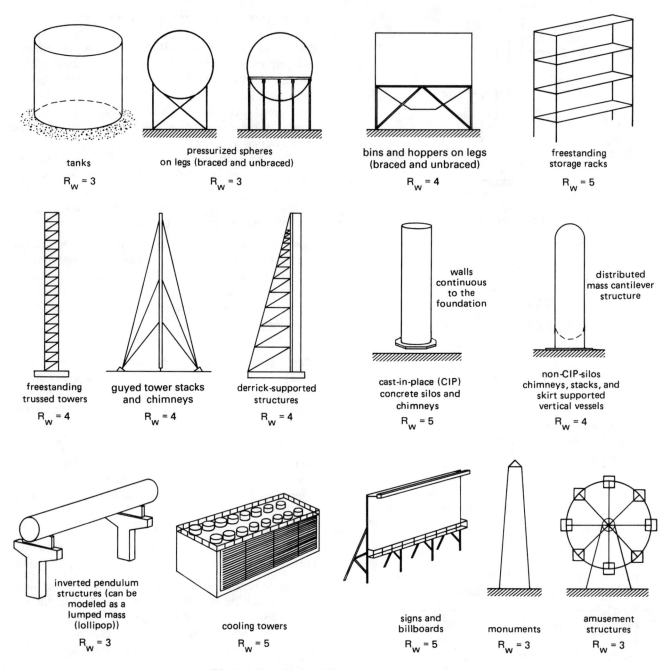

Figure 37 Typical Nonbuilding Structures

There are some differences, however, in how the base shear equations are used for building and nonbuilding structures. The natural period, T, must be determined from *method B* (i.e., a rational calculation, eigenvalue procedure, or dynamic analysis), and the 80% of the *method A* lower limit does not apply. (See Sec. 93 [UBC Sec. 2338(a)4].) The ratio C/R_w must be greater than 0.5 (as opposed to 0.075) [UBC Sec. 2338(d)1]. The maximum value of C is 2.75. Other provisions also apply.

The size of the supporting structural system for some short (i.e., less than 50 ft in height) nonbuilding structures is determined by the footprint of the structure, vibration limitation, or other operational considerations rather than traditional lateral loadings. In such cases, the support can be much stronger than required for seismic resistance, and it can be expected to remain in the elastic region during a maximum earthquake. Therefore, ductility is not a problem. The UBC [Sec. 2338(b)] permits these applications (with some restrictions) to use *intermediate moment resisting frames (IMRF)*. The value $R_w = 4$ (corresponding to low ductility demand during ground motion) must be used where this allowance is made [UBC Sec. 2338(b)].

Certain concrete pedestal-type structures can also be expected to remain in the elastic range during a maximum earthquake. However, the Blue Book commentary recommends that some ductility be included in the design. Such ductility is obtained (during the design stage) by providing sufficient transverse reinforcement to avoid brittle shear and/or development failures, and by providing continuity and development of longitudinal reinforcement.

The standard building drift limitations need not be met. However, the P-Δ effects must be evaluated if the drift limitations are not met [UBC Sec. 2338(a)5].

Table 16
UBC R_w Values for Nonbuilding Structures
[UBC Table 23-Q]

STRUCTURE TYPE	R_w
1. Tanks, vessels or pressurized spheres on braced or unbraced legs.	3
2. Cast-in-place concrete silos and chimneys having walls continuous to the foundation.	5
3. Distributed mass cantilever structures such as: stacks, chimneys, silos and skirt-supported vertical vessels.	4
4. Trussed towers (freestanding or guyed), guyed stacks and chimneys.	4
5. Inverted pendulum-type structures.	3
6. Cooling towers.	5
7. Bins and hoppers on braced or unbraced legs.	4
8. Storage racks.	5
9. Signs and billboards.	5
10. Amusement structures and monuments.	3
11. All other self-supporting structures not otherwise covered.	4

Reproduced from the 1991 edition of the
Uniform Building Code, copyright © 1991, with the
permission of the publishers, the International
Conference of Building Officials.

Example 13

A simple billboard-type sign is constructed on S_2 soil in seismic zone 4. It has a total weight of 3000 lbf distributed evenly across its width and 18 ft of height. Its total effective cross-sectional moment of inertia at the base of its supports is 0.05 ft^4. The modulus of elasticity of the three support posts is 2×10^6 psi. What is the UBC design base shear for an earthquake acting perpendicular to the sign face?

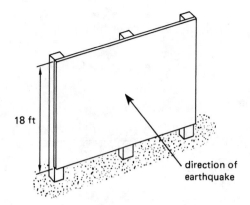

18 ft

direction of earthquake

Solution

One rational method of determining the sign's natural period is to consider the sign as an SDOF system. The supporting elements (i.e., the three posts) are essentially uniformly-loaded cantilevers. From standard beam deflection tables, the tip deflection, x, for such a configuration with distributed load w pounds per unit length is

$$x = \frac{wL^4}{8EI}$$

The stiffness, k, at the tip is

$$k = \frac{\text{total load}}{x} = \frac{wL}{x} = \frac{8EI}{L^3}$$

$$= \frac{(8) \left(2 \times 10^6 \, \frac{\text{lbf}}{\text{in}^2} \right) \left(144 \, \frac{\text{in}^2}{\text{ft}^2} \right) (0.05 \, \text{ft}^4)}{(18 \, \text{ft})^3}$$

$$= 19{,}750 \, \text{lbf/ft}$$

From Eqs. 29 and 30, the natural period of oscillation is

$$T = \frac{2\pi}{\omega} = 2\pi \sqrt{\frac{W}{kg}}$$

$$= 2\pi \sqrt{\frac{3000 \, \text{lbf}}{\left(19{,}750 \, \frac{\text{lbf}}{\text{ft}} \right) \left(32.2 \, \frac{\text{ft}}{\text{sec}^2} \right)}}$$

$$= 0.432 \, \text{sec}$$

This is greater than 0.06 s, so the billboard is not rigid as defined in UBC Sec. 2338(b).

From Table 13, the site coefficient for soil type S_2 is 1.2. From Eq. 77,

$$C = \frac{(1.25)S}{T^{\frac{2}{3}}} = \frac{(1.25)(1.2)}{(0.432)^{\frac{2}{3}}}$$

$$= 2.62 \quad (< 2.75, \, \text{ok})$$

From Table 16 for a sign or billboard, $R_w = 5$.

$$\frac{C}{R_w} = \frac{2.62}{5} = 0.52 \quad (> 0.50,\ \text{ok})$$

From Eq. 76,

$$V = \frac{ZICW}{R_w} = \frac{(0.4)(1.00)(2.62)(3000\ \text{lbf})}{5}$$
$$= 629\ \text{lbf}$$

◇   ◇   ◇   ◇   ◇   ◇   ◇

112 TANKS

Flat-bottom tanks and other tanks with supported bottoms at or below grade (as opposed to tanks on legs) are considered to be nonbuilding structures. Due primarily to the fact that liquid contents slosh around and add their own dynamic forces, the seismic performance of tanks is more complex than their simple appearance would suggest.[109]

The UBC [Sec. 2338(c)] allows three different types of analysis for tanks. First, the contents can be assumed to be rigid (i.e., have a period of less than or equal to 0.06 s) and the base shear calculated as $V = 0.5ZIW$ as described in Sec. 110. Second, a dynamic analysis that includes the inertial effects of the tank contents can be performed. Or, third, an approved national standard for design of tank supports can be used.[110]

Particular attention must be given to preventing uplift of tanks from their cradles or supports. Specifically, tanks must be anchored to their foundations. Furthermore, sloshing and freeboard should be considered in the design of the tank. Sloshing can be reduced by including baffles in the tank. Sufficient *freeboard* should be included in open tanks to prevent the contents from spilling out over the top.

113 ANCHOR BOLTS IN NONBUILDING STRUCTURES

Many nonbuilding structures are bolted to their foundations. In order to ensure ductile response, these bolts must stretch inelastically without failure and without being pulled from their concrete embedment. The bolt sizes (diameters and lengths) and placement patterns

should be chosen so that the bolts achieve their full strength in a maximum earthquake without failure. In particular, the Blue Book commentary recommends that bolt integrity be checked assuming a deformation of at least $3R_w/8$ times that expected from the design forces.

114 UBC DYNAMIC ANALYSIS PROCEDURE

Although the details of how to perform a dynamic analysis are beyond the scope of this book, the basic dynamic analysis procedure required by the UBC consists of three steps: (1) the static base shear is calculated; (2) a dynamic analysis using an elastic response spectrum is performed to determine the building period, base shear, story shears, and drifts; and (3) the results (with the exception of the period) are scaled upward in the ratio of the static to dynamic base shears [UBC Sec. 2335(e)3].[111]

The upward scaling provision of the UBC [Sec. 2335(e)3] is criticized by some structural engineers as being nonconservative. Depending on the site and geology, the scaling can double the forces for which the building must be designed.

Since the results are scaled, the magnitude of the design response spectrum is not as important as its shape (i.e., frequency content) and duration. Three different forcing inputs are permitted. (1) The UBC [Fig. 23-3] contains three different *standardized response spectra* (representing 5% damping) that can be used. (2) If available, site-specific design spectra from the actual building location can be used. (3) A time history analysis using accelerometer data from one or more actual earthquakes can be performed.[112] Enough modes must be included in the analysis to achieve a minimum 90% participation factor. [UBC Sec. 2335(e)] (See Sec. 64).

When performing a dynamic analysis, the minimum ground input must have a 10% (or greater) probability of occurring in a 50-year period [UBC Sec. 2335(b)]. This corresponds approximately to a Loma Prieta-sized earthquake. More significant events near 8.0 on the Richter scale (e.g., the 1906 San Francisco and the 1985 Mexico City earthquakes) are exceptional situations.

---

[109]Sloshing has very little damping (i.e., 0.1% or less).

[110]For example, American Petroleum Institute (API) publication 650, App. E, is an approved standard because it was developed to correspond to previous Blue Book provisions.

[111]Downward scaling is permitted but not required. However, the base shear cannot be scaled to less than 90% (for regular buildings) or 100% (for irregular buildings) of the static design value [UBC Sec. 2335(e)3].

[112]The incremental response of the structure should be digitized with a time step that is 3 to 10 times smaller than the shortest effective modal period.

Dynamic analysis is usually performed on a computer. However, the following steps can be used to carry out a manual dynamic analysis on a simple multistory structure when desired. (It is not practical to perform a dynamic analysis on structures with irregularities.)

step 1: Construct a lumped-mass, two-dimensional model of the structure. (i represents the mode index; x represents the floor index.)

step 2: Calculate the mode shape factors, $\phi_{i,m}$ (see Sec. 62). Normalize the mode shape factors so that $\phi = 1$ at the highest level.

step 3: Calculate the period, T_m, for each mode.

step 4: For each mode shape, calculate

$$L_m = \sum W_i \phi_{i,m} \qquad [78]$$

$$M_m = \sum W_i \phi_{i,m}^2 \qquad [79]$$

step 5: Calculate the seismic design coefficient, S_m, for each mode from the UBC normalized response spectra [UBC Fig. 23-3].

$$S_m = \frac{ZIC}{R_w} = (1.25)\left(\frac{ZIS}{R_w T^{\frac{2}{3}}}\right) \qquad [80]$$

step 6: Calculate the base shear for each mode.

$$V_m = \frac{L_m^2 S_m}{M_m} \qquad [81]$$

step 7: Calculate the participating mass fraction for each mode.

$$PM_m = \frac{L_m^2}{M_m W_t} \qquad [82]$$

$$W_t = \sum W_x \qquad [83]$$

step 8: Combine the base shears into the design dynamic lateral force, V_{dynamic}, using the SRSS (i.e., square root of the sum of the squares) method, with as many modes as are necessary to include at least 90% of the participating mass of the structural (i.e., until $\sum(PM) \geq 0.90$).

step 9: Calculate the lateral force, V_{static}, according to the static provisions of the UBC. Use Method A to determine the building period for the first mode.

step 10: Scale V_{dynamic} upward as is required by the UBC in Sec. 2335(e)3.

step 11: Distribute the scaled-up base shear to each level.

$$F_{x,m} = (V_{\text{dynamic}})\left[\frac{W_x \phi_{x,m}}{\sum(W_i \phi_{i,m})}\right] \qquad [84]$$

step 12: Determine the raw deflections, moments, and shears for each mode.

step 13: Use SRSS to combine the raw deflections, moments, and shears into effective values.

7

DIAPHRAGM THEORY

115 DIAPHRAGM ACTION

The story shears calculated by Eqs. 68 and 70 are assumed to be applied to a lumped mass representing the floor/ceiling layer in a building. The ceiling does not actually resist the story shear, but it does distribute the force among the resisting elements (e.g., shear walls, columns, moment-resisting frames, and other structural systems).

Ceilings and floors that transmit lateral forces to the resisting elements are known as *horizontal diaphragms*. The diaphragm's function of distributing the story shears is known as *diaphragm action*. It is common to refer to the story shear as the *diaphragm force*. However, it should be recognized that the diaphragm force may include some of the story shears (including the top force, F_t) for the level *and above* [UBC Sec. 2337(b)9B].

Plywood diaphragms are almost always considered flexible. Generally, concrete slab floors and diaphragms are considered to be rigid. Steel deck diaphragms and poured-in-place gypsum floors can be either rigid or flexible, depending on their design.

116 SEISMIC WALL AND DIAPHRAGM FORCES

Figure 38 shows a simple (regular) one-story box building with a flexible diaphragm roof. (While this discussion is applicable to larger buildings, Secs. 116 through 133 are primarily concerned with simple one-story buildings with masonry or reinforced concrete walls.) Both walls are identical. An earthquake acceleration occurs in the direction shown by the ground motion arrow.

When discussing seismic forces in structures with diaphragms (e.g., one- or two-story masonry buildings

with plywood diaphragm floors and ceilings), it is important to distinguish between forces in the parallel and perpendicular walls. (See Fig. 38.) The forces in the parallel walls are shear forces, while the forces in the perpendicular walls are normal forces (i.e., compressive and tensile forces). This section is primarily concerned with the shear force in the parallel walls. Forces in the perpendicular walls (i.e., the *chord forces*) are covered in Sec. 131.

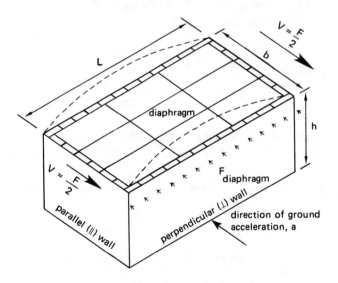

Figure 38 Simplified Building and Roof Diaphragm

The seismic shear force acting on the parallel walls depends on the mass being accelerated, which consists of the diaphragm weight and some portion, usually assumed to be half, of the total wall weight. (It is assumed that the seismic shear force from the remaining half of the total wall weight passes directly to the foundation without stressing the wall.) The diaphragm weight includes the weight of any equipment mounted on the roof, anything suspended inside the building from the roof, and anything mounted on the upper half of the walls. The wall weight includes all weight of any parapet that projects above the roof line.

Forces are calculated either as $F = ma = (W/g)a$ or from the UBC equation (See Eq. 73).[113] Both depend on the weight, W, of the structure. (See Sec. 97.) Openings, such as windows and doors, in the walls that reduce the wall weight are usually disregarded when determining wall weight.

The total seismic force resisted by the two parallel walls near the ground level is the sum of seismic forces resulting from the diaphragm and wall weights. In the simple illustration of Fig. 38, the force on one wall is half of the total force for both rigid and flexible diaphragms.

The portion of seismic load originating from the acceleration of the perpendicular walls is given by Eq. 85. The symbols $\perp$ and $\parallel$ refer to "perpendicular" and "parallel," respectively.

$$F_{\perp\text{walls}} = \left(\frac{1}{2}\right) ZIC_p W_{\perp\text{walls}}$$
$$= \left[\left(\frac{1}{2g}\right) W_{\perp\text{walls}}\right] a \qquad [85]$$

The portion of seismic load originating from the acceleration of the parallel walls is

$$F_{\parallel\text{walls}} = \left(\frac{1}{2}\right) ZIC_p W_{\parallel\text{walls}}$$
$$= \left[\left(\frac{1}{2g}\right) W_{\parallel\text{walls}}\right] a \qquad [86]$$

Equation 87 calculates the seismic force on the diaphragm at the diaphragm-to-parallel wall connection.

$$F_{\text{diaphragm}} = F_{\perp\text{walls}} + ZIC_p W_{\text{diaphragm}} \qquad [87(\text{a})]$$

$$F_{\text{diaphragm}} = F_{\perp\text{walls}} + \left[\left(\frac{1}{g}\right) W_{\text{diaphragm}}\right] a \qquad [87(\text{b})]$$

Equation 88 calculates the total shear force on the parallel walls.

$$F_{\text{total}} = F_{\text{diaphragm}} + F_{\parallel\text{walls}} \qquad [88]$$

There are two reasons for calculating the forces from the diaphragm and parallel walls separately. The first reason is to distinguish between the two for the purpose of subsequent calculations; that is, the parallel wall force

does not contribute to chord loads and diaphragm shear where the diaphragm is flexible. The second reason is to emphasize the timing difference that occurs in a real earthquake.

The perpendicular and parallel walls experience an almost immediate force due to ground acceleration. However, the parallel walls receive the diaphragm force only after some delay. Unfortunately, an accurate analysis of this aspect of seismic behavior is almost impossible. For simple structures with three or fewer floors, the static method of adding all forces together must suffice.

As with any seismic analysis, the diaphragm force must be evaluated in both orthogonal directions.

117 WALL SHEAR STRESS

In the simple building shown in Fig. 38, the rigidities (for a rigid diaphragm) and tributary areas (for a flexible diaphragm) are identical for the two walls. Therefore, half of the total seismic force is carried by each parallel wall. The shear stress, v, in a parallel wall of thickness t is[114]

$$v_{\text{total}} = \frac{F_{\text{total}}}{2b} \qquad \text{[per unit length]} \qquad [89(\text{a})]$$

$$v_{\text{total}} = \frac{F_{\text{total}}}{2bt} \qquad \text{[per unit area]} \qquad [89(\text{b})]$$

Shear walls located on adjoining levels should be structurally continuous and not be offset. There should be a complete transmission path from a shear wall on one level to another shear wall below.

Horizontally- and vertically-stacked openings in shear walls need special attention. Vertical shears need to be transferred to adjacent piers or boundary columns.

---

[113]Equation 62 cannot be used to calculate the seismic force on a wall, diaphragm, or subdiaphragm, which are *parts* of the structure. Equation 73 (for parts and portions) must be used.

[114]The allowable shear stress in a masonry wall depends on the compressive strength of the masonry, f'_m. When in-plane flexural reinforcement is present, the maximum allowable shear stress is 35 psi. Similarly, with shear reinforcement (i.e., the horizontal steel) taking all the shear, the maximum shear stress is 75 psi [UBC Sec. 2406(c)7B]. The allowable shear stress on reinforced walls without special inspection is always half of what is permitted with special inspection [UBC Sec. 2406(c)1]. Horizontal steel must be provided to carry all shear stress higher than the allowable value. While the horizontal steel is used for shear, it is normal practice to have the same reinforcement ratio horizontally and vertically. Unreinforced masonry walls are not permitted in seismic zones 3 and 4 [UBC Sec. 2407(h)4B]. Therefore, the empirical design method [UBC Sec. 2407(i)] and UBC Table 24-H ("Allowable Compressive Stresses for Empirical Design in Masonry") cannot be used in those zones.

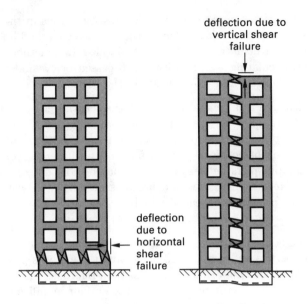

Figure 39 Stacked Openings

It should be noted that the UBC [Sec. 2407(h)4F(i)] requires masonry shear walls in seismic zones 3 and 4 to be designed to resist 150% of the force calculated from Eq. 63.

118 RIGID DIAPHRAGM ACTION

A *rigid diaphragm* does not change its plan shape when subjected to lateral loads. It remains the same size, and square corners remain square. There is no internal bending. The same deflection (i.e., drift) is experienced by all parts of the diaphragm. Rigid diaphragms are capable of transmitting torsion to the major resisting elements (usually the outermost elements). The lateral story shear is distributed to the resisting elements in proportion to the rigidities of those elements.

Figure 40 illustrates a simple arrangement of a rigid diaphragm distributing the seismic load to two shear walls.

In seismic zones 3 and 4, the shear stress must not exceed the limits given in the UBC for masonry shear walls [Sec. 2407(h)4]. (See Ftn. 114.) Since the diaphragm force is distributed to the resisting elements in proportion to the rigidities of those elements, the rigidities must be determined. In practice, a few guidelines are needed to do so.

1. The relative rigidities of masonry or concrete structures can be calculated using Eqs. 26 and 27. Alternatively, Apps. D and E can be used. There is no need to use actual values of E and G, since only relative values are needed.

2. If a wall extends above roof level (i.e., has a *parapet*), the distance above the roof (i.e., the parapet height) should be disregarded when calculating the rigidity.

3. Shear walls with openings such as doors and windows require special attention. As a first approximation, such a wall can be treated as a solid shear wall. However, other methods (see Sec. 119) exist for evaluating the overall wall rigidity.

4. The rigidities of *transverse walls* (i.e., walls running perpendicular to the direction of the lateral force) are usually disregarded for calculating direct loads. This is called "omitting the *weak walls*." However, rigidities of all walls must be known in order to calculate torsional loads.

◇   ◇   ◇   ◇   ◇   ◇   ◇

Example 14

Two walls—wall A with rigidity of 3.278 and wall B with rigidity of 7.895—support a rigid diaphragm roof in a one-story reinforced masonry building. A total seismic force of 120,000 lbf is applied along the diaphragm centerline. Determine the shear carried by each wall.

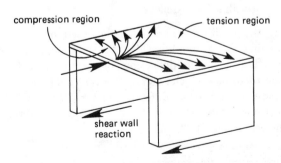

Figure 40 Rigid Diaphragm Action

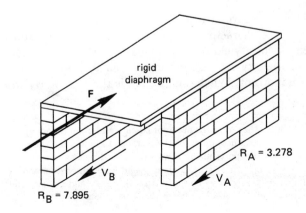

Solution

The total seismic force is distributed to the walls in proportion to the rigidities.

$$V_A = \left(\frac{R_A}{R_A + R_B} \right) F$$

$$= \left(\frac{3.278}{3.278 + 7.895} \right) (120{,}000 \text{ lbf})$$

$$= 35{,}206 \text{ lbf}$$

$$V_B = \left(\frac{R_B}{R_A + R_B} \right) F$$

$$= \left(\frac{7.895}{3.278 + 7.895} \right) (120{,}000 \text{ lbf})$$

$$= 84{,}794 \text{ lbf}$$

◇   ◇   ◇   ◇   ◇   ◇   ◇

119 CALCULATING WALL RIGIDITY

In order to determine the rigidity of a wall with openings, it is necessary to divide the wall into piers and beams. A *pier* is a vertical portion of the wall whose height is taken as the smaller of the heights of the openings on either side of it. A *beam* is a horizontal portion left after the piers have been located.

Figure 41 illustrates a wall with two windows. $P_1, P_2,$ and P_3 are piers. B_1 and B_2 are beams.

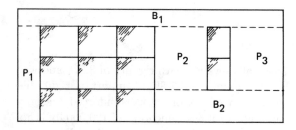

Figure 41   A Wall with Openings

There are several methods of calculating wall rigidity from the characteristics of the wall's piers and beams, all of which yield slightly different answers.[115] The actual rigidity for any particular wall in a structure is unimportant. What is important is the *relative rigidity*

---

[115]These other methods are approximate and often do not agree. Also, it is not uncommon for the methods to determine the rigidity of a wall—particularly a wall with fixed-pier assumptions—with openings to be greater than a solid wall (an obvious impossibility). The rigidity of a wall with openings should be compared to the rigidity of a solid wall of the same dimensions.

of the wall compared to all other resisting walls in the structure. Thus, it is essential that the same method be used to calculate the rigidities of all walls. Two methods for calculating wall rigidity are discussed below under "Method A" and "Method B" in this section.

Although it is generally assumed that cantilever conditions (see Sec. 48) prevail for the walls in one- and two-story buildings taken in their entireties, piers within the wall can be considered either fixed or cantilevered. For example, piers between openings may be considered to be fixed at their tops and bottoms although the wall taken as a whole is cantilevered.

The accuracy in wall rigidity calculations is not great. There are many assumptions made about material properties and wall performance, and the analysis procedure, though formalized, is less than rigorous. Therefore, values with more than three or four significant digits are unwarranted.

Method A

With this fast and simple method (used only for preliminary analyses), the rigidity of a wall is calculated as the sum of the rigidities of the individual piers framed between openings in a wall. All piers are assumed to be fixed. The pier height is the height of the shortest adjacent opening.[116] Beams and wall portions above and below the openings are not considered.

Method B

By far the most commonly used method of evaluating wall rigidities calculates "deflections" from standardized values of force, thickness, and modulus of elasticity. These deflections are recognized as being the reciprocals of rigidity.

To start, the gross deflection of the solid wall is calculated, ignoring all openings and assuming cantilever action. Then the strip deflection of an interior strip having length equal to the wall length and height equal to the tallest opening is calculated, again assuming cantilever action. This strip deflection is subtracted from the solid wall's gross deflection.

Next, the rigidities (not the deflections) of all piers (assuming fixed ends) within the removed strip are summed, and the pier deflection correction is calculated as the reciprocal of the sum. The pier deflection correction is added to the difference of the gross and strip deflections to give the net deflection. The wall rigidity is the reciprocal of this net deflection.

---

[116]A common error is to use the ground-to-ceiling distance.

The rigidity of the wall must be calculated by this method one opening at a time, considering the fixed pier adjacent to the opening and the wall section below the opening. The calculation of the pier deflection correction becomes recursive when openings in the wall are of different heights. For this reason, Method B can take a long time if the wall is relatively complex.

Example 15

The masonry wall shown in Fig. 41 and dimensioned below has a uniform thickness and is part of a one-story building. Determine the rigidity using the two methods described in Sec. 119.

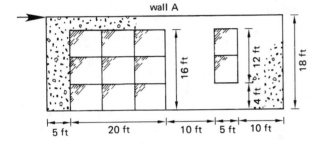

Solution

Method A

The total rigidity of the wall is the sum of all the pier rigidities. Beams are disregarded.

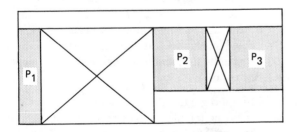

The rigidities, R, of all piers (assumed fixed) are obtained from Eq. 26 or App. D after calculating the height/depth ratio, h/d, of each.

element	h	d	h/d	R
P_1	16	5	3.2	0.236
P_2	12	10	1.2	1.877
P_3	12	10	1.2	1.877

In this method, the rigidity of the wall is the sum of the individual fixed-pier rigidities.

$$R = 0.236 + 1.877 + 1.877 = 3.99$$

Method B

First determine the deflection of a solid wall. The height/depth ratio of the entire wall is

$$\frac{h}{d} = \frac{18\,\text{ft}}{50\,\text{ft}} = 0.36$$

Since this is a one-story building, the wall taken as a whole is assumed to be cantilevered. From App. E, the rigidity is 7.895. The "deflection" of the solid wall is

$$\Delta_\text{solid} = \frac{1}{R} = \frac{1}{7.895} = 0.1267$$

The highest opening has a height of 16 ft, so a "mid-strip" 16 ft high and 50 ft long is removed. Assuming a solid wall, the height/depth ratio is

$$\frac{h}{d} = \frac{16}{50} = 0.32$$

The entire wall was assumed to be a cantilever pier, and this mid-strip represents the majority of the wall. Therefore, it also is a cantilever member. Appendix E gives the rigidity as 9.165. The "deflection" of an assumed solid mid-strip is

$$\Delta_\text{mid-strip} = \frac{1}{9.165} = 0.1091$$

The mid-strip, however, is not solid. It consists of two functional parts: a 5 ft by 16 ft solid pier (P_1) at the left and a larger section with a small window at the right. The large window section running floor-to-ceiling is assumed to contribute no rigidity to the wall. The rigidity of this mid-strip is the sum of the rigidities of pier P_1 and the larger section.

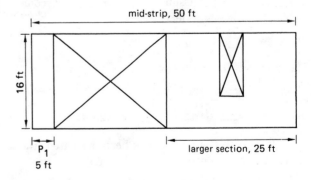

For the solid pier, P_1, at the left,

$$\frac{h}{d} = \frac{16}{5} = 3.2$$

For the same reason for considering the mid-strip to be a cantilever, this pier is assumed to act as a cantilever. From App. E, $R = 0.071$. (The "deflection" of pier P_1 is not needed.)

The larger section at the right of the wall has dimensions of 16 ft by 25 ft. Taken as a solid cantilever pier, the gross rigidity and deflection are

$$\frac{h}{d} = \frac{16}{25} = 0.64$$

$$R_{\text{larger section, gross}} = 3.369$$

$$\Delta_{\text{larger section, gross}} = \frac{1}{3.369} = 0.2968$$

However, the larger section is not itself solid. There is a 12-ft-high window. For a 12 ft by 25 ft solid window strip assumed to act as a cantilever,

$$\frac{h}{d} = \frac{12}{25} = 0.48$$

$$R_{\text{window strip}} = 5.313$$

$$\Delta_{\text{window strip}} = \frac{1}{5.313} = 0.1882$$

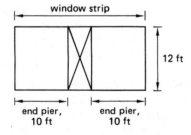

The window strip contributes stiffness, as there are two 12 ft by 10 ft end piers. Although cantilever performance could be argued as well, these two end piers are assumed to be fixed. Their combined rigidity and deflection are

$$\frac{h}{d} = \frac{12}{10} = 1.2$$

$$R_{\text{end piers}} = 2 \times 1.877 = 3.754$$

$$\Delta_{\text{end piers}} = \frac{1}{3.754} = 0.2664$$

Now that all the pieces have been evaluated, the rigidity of the entire wall can be built up.

The net deflection and net rigidity of the larger section is

$$\Delta_{\text{larger section, net}} = 0.2968 - 0.1882 + 0.2664$$
$$= 0.3750$$

$$R_{\text{larger section, net}} = \frac{1}{0.3750} = 2.667$$

(Check that 2.667 is less than the gross value of 3.369.)

Since the highest opening in the mid-strip extends the full height, the rigidity is merely the sum of the rigidities of pier P_1 and the larger section.

$$R_{\text{mid-strip, net}} = 0.071 + 2.667 = 2.738$$

$$\Delta_{\text{mid-strip, net}} = \frac{1}{2.738} = 0.3652$$

(Check that 2.738 is less than the gross value of 9.165.)

The deflection and relative rigidity of the entire wall is

$$\Delta_{\text{entire wall}} = 0.1267 - 0.1091 + 0.3652 = 0.3828$$

$$R_{\text{entire wall}} = \frac{1}{0.3828} = 2.612$$

(Check that 2.612 is less than the gross value of 7.895.)

120 FLEXIBLE DIAPHRAGMS

A flexible diaphragm changes shape when subjected to lateral loads. Its forward edges bend outward, and its back edges bend inward, with a deflection shape similar to that of a simply-supported beam loaded uniformly. Flexible diaphragms are assumed to be incapable of transmitting torsion to the resisting elements. (Also, see Ftn. 64.)

A flexible diaphragm distributes the diaphragm force in proportion to the tributary areas of the diaphragm, as opposed to distributing it in proportion to the rigidities of the vertical resisting elements, as does a rigid diaphragm.

As defined by the UBC, a *flexible diaphragm* is one that has a maximum lateral deflection more than two times the average story drift [UBC Sec. 2334(f)]. To determine if a diaphragm is flexible, compare the in-plane deflection at the midpoint of the diaphragm to the story drift of the adjoining vertical resisting elements under equivalent tributary load.

This definition does not mean that the diaphragm and the vertical resisting elements to which the diaphragm is connected become disconnected in an earthquake. Obviously, they should not, and the deflection of vertical elements and the diaphragm will be the same. However, the criterion is tested by comparing the deflection the diaphragm would experience under a fixed loading to the deflection the vertical resisting elements would experience under the same loading.

121 FRAMING TERMINOLOGY

Figure 42 illustrates such common framing terms as *sheathing, girder, beam, purlin,* and *joist* (or *sub-purlin*), as well as the *bridging* and *blocking* that are used to prevent lateral buckling. Usually blocking (e.g., often cut from the same material as the joists, although other blocking techniques are used) frames into joists or sub-purlins (two-by-sixes, two-by-eights, etc.); joists frame into purlins (e.g., four-by-eights); purlins frame into beams (e.g., four-by-fourteens); beams frame into girders (e.g., glulams); and girders frame into the walls.[117]

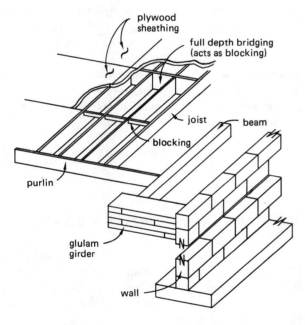

Figure 42 Framing Members

122 DRAG STRUTS

It is required that *drag struts* (also known as *collectors, braces,* or merely *struts* or *ties*) be used to transmit diaphragm loads between chords at points of discontinuity (irregularity) in the plan [UBC Sec. 2337(b)9C]. These

---

[117]These terms are not so rigidly defined that they preclude incorrect usage.

drag struts effectively separate the diaphragm into sub-diaphragms that can be analyzed independently.

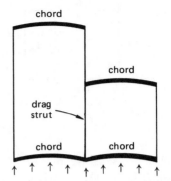

Figure 43 Use of a Drag Strut

Diaphragm sheathing alone may not be relied on to transfer loads between chords [UBC Sec. 2337(b)9D].

123 DRAG FORCE

The drag force is the product of the diaphragm load (per unit area) and the areas tributary to the drag strut. Tributary areas are usually taken as some fraction of the sub-diaphragm areas located on both sides of the strut. The force is not the same along the length of the drag strut but increases to a maximum at the point where the drag strut frames into a parallel wall chord.

124 COLUMNS SUPPORTING PARTS OF FLEXIBLE DIAPHRAGMS

An interior or exterior column can be used to support the vertical roof or floor load, but such a column usually provides no lateral support when the diaphragm is flexible. A girder that frames into such a column at one girder end and a shear wall at the other girder end almost always acts as a collector (strut) for seismic forces parallel to the girder direction.

Example 16

The plan view of an irregular building is shown. (a) Determine the tributary areas for an earthquake in the north-south direction. (b) Determine where the force in the drag strut is maximum.

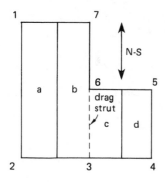

Solution

(a) The walls along sides 1-2, 6-7, and 4-5 will resist the seismic force. (Walls 1-7, 5-6, and 2-3-4 are *weak walls*. See Sec. 118.) The drag strut between areas b and c splits the diaphragm into two rectangular subdiaphragms: a-b and c-d.

Area a is tributary to wall 1-2. Area b is tributary to wall 6-7. The forces transmitted to the drag strut are carried by the drag strut back to wall 6-7.

Area d is tributary to wall 4-5. Area c is tributary to the drag strut, which transmits all of the area c diaphragm force into wall 6-7.

(b) The drag member carries half of area b's diaphragm force and all of area c's diaphragm force. The maximum value of this force occurs at point 6, where the drag member frames into the shear wall 6-7.

125 FLEXIBLE DIAPHRAGM CONSTRUCTION

A flexible diaphragm is a relatively thin structural element such as a roof or floor attached to relatively rigid walls. It can be constructed as a braced frame with non-structural covering, or as joists sheathed with plywood, boards, insulating board, or gypsum sheets.

Rectangular diaphragms are most common and are the simplest to analyze. However, diaphragms do not need to be flat in order to resist shear. A plywood roof can be inclined, peaked (i.e., folded-plate), or curved. In such cases, the roof trusses act as web stiffeners. A non-flat diaphragm is analyzed according to its footprint (plan) dimensions.

There are three structural requirements imposed on flexible diaphragms:

1. The diaphragm must be strong enough to remain intact under the action of wind and seismic loads.

2. The diaphragm must be securely attached to a wall in order to resist forces parallel to the wall.

3. The diaphragm must be securely attached to a wall in order to resist forces perpendicular to the wall.

126 FLEXIBLE DIAPHRAGM TORSION

There is no torsional shear stress (see Sec. 76) from eccentric mass placement in either the walls or diaphragm because flexible diaphragms are not capable of distributing torsional shear stresses.

127 DIAPHRAGM SHEAR STRESS

In most cases, the criterion by which diaphragm construction and connections is evaluated[118] is the *diaphragm shear stress*, $v_{\text{diaphragm}}$.[119] The stress is assumed to exist uniformly across the length b, known as the *diaphragm depth*. (The total force is shared by the two parallel walls, each of depth b. The perpendicular walls do not resist the applied seismic force.) The parallel wall force is not included in the diaphragm shear loading.

$$v_{\text{diaphragm}} = \frac{F_{\text{diaphragm}}}{2b} \quad \text{[per unit length]} \quad \text{[90]}$$

128 DIAPHRAGM NORMAL STRESS

A flexible diaphragm is designed to withstand shear in its plane. It has no bending strength of its own. Rather, the diaphragm relies on the stiffness of the perpendicular walls to limit overall diaphragm deflection. A common analogy is to assume the diaphragm acts like a girder, where the flanges (i.e., the perpendicular walls) resist the bending moment, and the web (i.e.,

---

[118]The larger of the seismic and wind shear stresses will be used in the design, but not both.

[119]The maximum allowable shear stress on plywood diaphragms depends on the nail size and spacing, plywood thickness, and width of framing members, and on whether or not the plywood edges are blocked, among other factors. Typical values range from 100 to 800 pounds per foot, with the lower values (i.e., 100 to 300 pounds per foot) applying to unblocked diaphragms. (See Table 17.) Refer to the code for exact shear stress limits and required nail spacing [UBC Tables 25-J-1 and 25-J-2].

the diaphragm) resists the shear. This is illustrated in Fig. 44.

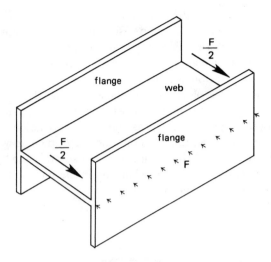

Figure 44   Web and Flange of a Girder

The diaphragm shear stress is assumed to be linearly distributed from zero at the midpoint (i.e., at $L/2$) to $v_{diaphragm}$ at the parallel walls. (Distance L in Fig. 45 is known as the *diaphragm span*). At any particular point, the shear stress is uniform across the diaphragm between perpendicular walls. (The girder analogy fails here since the shear stress distribution between perpendicular walls is not parabolic.) Edge nailing of the plywood sheets to the blocking keeps the shear resistance continuous across the diaphragm.

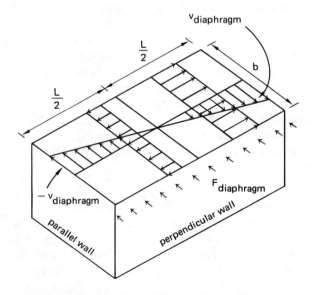

Figure 45   Shear Stress Distribution
on a Diaphragm

129 DEFLECTION OF FLEXIBLE DIAPHRAGMS

A flexible diaphragm cannot resist deflection. The beam analogy described in Sec. 128 is also valid here, as the diaphragm assumes the deflection shape of a simply-supported beam loaded by a uniform load.[120] The deflection is resisted by the perpendicular walls, which, being masonry or concrete, are less tolerant of compression and tension. Since the perpendicular wall deflection is the same as the diaphragm deflection, this deflection may be the factor that limits the force that can be safely applied to the diaphragm.

Actual determination of the deflection in a plywood diaphragm is complex, as it is for any wood/timber structural member, but procedures are available. The following equation (reference: UBC Standard 25-9) calculates the plywood diaphragm deflection as a sum of *flexural distortion* (the first term), *shear distortion* (the second term), and *nail distortion* (the third term). In some cases, such as in wood-framed perpendicular walls where a wood double-plate serves as the chord, the fourth term may be added to account for *chord-splice slip* values. Such slippage is neglected with masonry walls.

$$\text{deflection (in)} = \frac{5vL^3}{8EAb} + \frac{vL}{4Gt}$$
$$+ 0.188Le_n + \frac{\sum(\Delta_c x)}{2b} \quad [91]$$

v = maximum shear in pounds per foot (plf)

L = diaphragm length (feet)

E = modulus of elasticity of chord (psi)

b = diaphragm width (feet)

A = section area of chord (square inches)

t = plywood thickness (inches)

e_n = nail slip, at load per nail (inches)

$\sum(\Delta_c x)$ = sum of individual chord-splice slip values on both sides of the diaphragm, each multiplied by its distance to the nearest support

G = plywood modulus of rigidity (psi, typically taken as 90,000 psi for structural plywood. Alternatively, $G = E/20$ for panels with exterior glue.)

The nail slip, e_n, depends on the nail size, plywood thickness, load per nail, and type of lumber. Usually,

---

[120]Another analogy is that the diaphragm hangs like a wet sheet from the walls.

worst-case green lumber is assumed. Nail slip is usually obtained graphically from appropriate sources. Values are typically less than 0.15 in, with most values being half that amount.

The maximum deflection of the diaphragm is the acceptable limitation of deflection or drift for the perpendicular walls directly below the diaphragm. This limitation will depend on whether the walls are masonry or concrete and on which authority specifies the limitation. The UBC does not specify actual limitations on diaphragm deflection, but it limits such deflection to amounts that maintain the structural integrity and protect occupants [UBC Sec. 2337(b)9A].[121]

Diaphragm deflection calculations are generally unnecessary and are waived if the UBC provisions (Table 25-I) for maximum *diaphragm ratios* for size are followed. (See Sec. 144.) The span-to-width (L/b) ratio for most horizontal diaphragms (including edge-nailed plywood diaphragms) is limited to 4:1, except for conventionally constructed diagonal sheathing (as defined by the UBC in Sec. 2513(b)) for which the limitation is reduced to 3:1.

If the deflection is excessive, it can be reduced by increasing the plywood thickness, decreasing the nail spacing, adding a collector strut, or placing an additional shear wall within the building to reduce the diaphragm span.

130 CHORDS

The elements—walls or reinforcement—capable of supporting normal (i.e., compressive and tensile) forces at the edges of the diaphragm along the perpendicular walls are known as *chords*. Chords are generally considered to be tension and compression members, analogous to the flanges of the beam shown in Fig. 44. A diaphragm will be constructed with chord elements along all outer edges. The chords that run perpendicular to the applied force, that is, along the perpendicular walls, and are stressed during an earthquake are called the *active chords*. The chord elements in parallel walls are known as the *passive chords*.

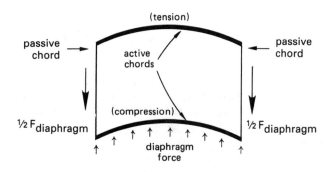

Figure 46   Chords

[121]The California *Office of Architecture and Construction* limits deflection to 1/16 in per foot of wall height. SEAOC has developed the following formula for the maximum allowable deflection:

$$\text{deflection (in)} = \frac{75h^2 F_b}{Et}$$

In the above equation, h is the wall height (feet), F_b is the allowable masonry flexural (compressible) stress (psi, increased by the 1/3 allowance for seismic loads), E is the masonry modulus of elasticity (1,500,000 psi for masonry, 2,000,000 psi for concrete), and t is the wall thickness (inches).

A similar equation has been proposed by the American Institute of Timber Construction, in which the 75 is replaced by 96.

The *Reinforced Masonry Engineering Handbook* (see App. G) suggests the following equation for maximum deflection. This equation can be derived from the SEAOC equation if $E = 1{,}500{,}000$ psi and $F_b = (4/3)(900 \text{ psi})$ is used. 900 psi is appropriate for concrete walls but may be too high for masonry walls.

$$\text{deflection (in)} = \frac{2h^2}{45t}$$

In masonry buildings, the perpendicular walls themselves can (and, are intended to) serve as chords if the force is transferred through solidly attached *ledgers* (see Sec. 149). This assumes that the masonry walls have adequate tensile reinforcement. Alternatively, a properly spliced wood ledger beam or a bond beam using an embedded reinforcing bar could serve as the chord with masonry walls, depending on the method of attachment. Chords can also be wood (e.g., the double top-plate in conventional wood stud walls), steel, or any other continuous material connected to the diaphragm edge.

The diaphragm may be functionally divided into independent parts, known as *sub-diaphragms*. In this case, chords and struts will run through the diaphragm in addition to around it [UBC Sec. 2337(b)9C]. One method of supporting internal chord members in masonry construction is with pilasters, as shown in Fig. 47. A *pilaster* is constructed as part of the masonry wall and is designed as a column.

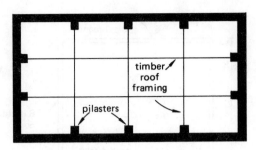

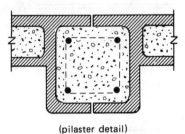

(pilaster detail)

Figure 47 A Pilaster

131 CHORD FORCE

The diaphragm itself is assumed incapable of supporting a normal (bending) stress. Chords are designed to carry all tensile and compressive forces in their respective walls.

The maximum chord force occurs at mid-span ($L/2$). The maximum chord force, C, is calculated as the bending moment of a simple beam under a distributed load (i.e., $wL^2/8$, with w in plf) divided by the depth, b, of the diaphragm. The distributed load is $w = F_{\text{diaphragm}}/L$, where $F_{\text{diaphragm}}$ is in pounds. (For wind loads, w is the wind load alone.)

$$C = \frac{M}{b} = \frac{wL^2}{8b} = \frac{F_{\text{diaphragm}}L}{8b} \quad [92]$$

Equation 92 is derived by equating the applied moment to the resisting moment. The applied moment is $wL^2/8$. The resisting moment comes from two sources: the two perpendicular walls. One perpendicular wall is in compression; the other is in tension. Both act with a moment arm of $b/2$ (with respect to a neutral axis passing through the midpoint of the diaphragm).

$$\frac{wL^2}{8} = C\left(\frac{b}{2}\right) + C\left(\frac{b}{2}\right) = Cb \quad [93]$$

$$C = \frac{wL^2}{8b} \quad [94]$$

The minimum chord force is zero and occurs at the chord ends. At intermediate locations, the force follows the shape of the bending moment between the two parallel walls.

132 CHORD SIZE

The required chord size (cross-sectional area) can be determined from the allowable stress in tension or compression, whichever is less, for the chord material. A one-third increase is permitted because the loading is seismic [UBC Sec. 2504(c)4ii].

$$A_{\text{chord}} = \frac{C}{1.33 \times \text{allowable stress}} \quad [95]$$

The chord area may be reduced near the parallel walls, but the reduction must be in accordance with the actual moment distribution.

The allowable tensile stress for steel reinforcing bar depends on its grade (equivalent to its minimum yield strength in ksi). Allowable tensile stresses are 20 ksi for grades 40 and 50 steel, and 24 ksi for grade 60 steel [UBC Sec. 2626(d)2A and B].

133 OVERTURNING MOMENT

In addition to seismic shear loading, a parallel wall (see Fig. 38) will be subjected to overturning moments as well. Overturning will not be a problem, however, if there is a larger resisting moment.

One of the components contributing to overturning is the seismic force of $v_{\text{diaphragm}}$ per unit length along the top of the parallel wall. If the parallel wall has a length of b', the total overturning force parallel to the ground is $v_{\text{diaphragm}}b'$. (b' and b may be the same if the parallel wall is one piece, or b' may correspond to the length of a tilt-up wall section.) This total roof load acts with a moment arm of h, the height of the force above the ground.

Also contributing to overturning is the seismic force on the parallel wall, calculated as $W_{\text{wall}}a/g$ or ZIC_pW_{wall}. This seismic force acts halfway up the parallel wall, with a moment arm of $h/2$.

$$M_{\text{overturning}} = v_{\text{diaphragm}}b'h + \frac{W_{\text{wall}}ah}{2g}$$
$$= v_{\text{diaphragm}}b'h + \frac{ZIC_pW_{\text{wall}}h}{2} \quad [96]$$

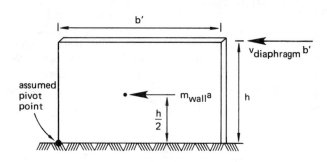

Figure 48 Overturning Forces on a Parallel Wall

The overturning moment is resisted by the weight of the parallel wall and the distributed roof dead load (calculated as the roof load tributary to that panel), both acting with a moment arm of $b'/2$.

$$M_{\text{resisting}} = (\text{roof dead load} + W_{\text{wall}})\left(\frac{b'}{2}\right) \quad [97]$$

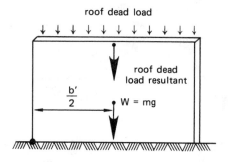

Figure 49 Resisting Forces on a Parallel Wall

Measures to resist overturning are well known and include anchoring the parallel wall panels to the foundations and, in the case of tilt-up slab construction, interconnecting adjacent panels.

Example 17

A simple four-walled 40 ft by 50 ft building is part of a hazardous chemical chlorine storage facility located in seismic zone 4. It is constructed with a plywood-sheathed roof on 10-ft-high concrete masonry unit (CMU) bearing walls 8 in thick. All walls are reinforced vertically and horizontally. The average weight of the roof diaphragm and mounted equipment is 20 psf (pounds per square foot). All connections between walls, roof, and foundation are pinned. The building performance is being analyzed for an earthquake acting parallel to the short dimension. Disregard all openings in the walls.

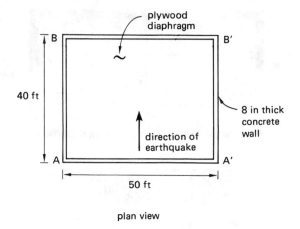

plan view

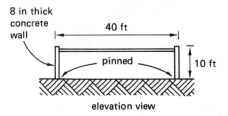

elevation view

(a) Find the diaphragm-to-wall shear (per foot of wall) on line A-B.

(b) Find the required diaphragm edge nailing spacing on line A-B. (Assume 6d nails, case 1 plywood layout, 3/8 in structural II plywood on blocked 2 in frame members.)

(c) Find the maximum chord force on line B-B'.

(d) Find the horizontal shear stress (in psi) in wall A-B at a point 5 ft above the foundation.

(e) Determine whether the wall thickness is adequate.

(f) If wall A-B was to be reduced to only 10 ft long and the remaining 30 ft of roof supported by a collector, find the collector force at the end of the wall.

Solution

From Table 11, the seismic zone factor for seismic zone 4 is $Z = 0.40$. From Table 12 for a hazardous facility, the importance factor, I, is 1.25. Inasmuch as the soil type is unknown, the maximum value of C (2.75) must be used. (See Sec. 93.)

From Table 14 for a bearing wall system consisting of concrete shear walls, $R_w = 6$. Thus, the base shear equation (Eq. 62) is

$$V = \frac{ZICW}{R_w} = \frac{(0.40)(1.25)(2.75)W}{6}$$
$$= 0.229W$$

The weight being accelerated by the earthquake consists of the diaphragm weight and a portion of the wall weight. The weight of the diaphragm is

$$W_{\text{diaphragm}} = \left(20 \, \frac{\text{lbf}}{\text{ft}^2}\right)(40 \, \text{ft})(50 \, \text{ft})$$
$$= 40,000 \, \text{lbf}$$

Since concrete has a density of 150 pcf (pounds per cubic foot), the weight of 1 ft^2 of an 8-in-thick wall is

$$\gamma = \frac{(8 \, \text{in})\left(150 \, \frac{\text{lbf}}{\text{ft}^3}\right)}{12 \, \frac{\text{in}}{\text{ft}}}$$
$$= 100 \, \text{lbf/ft}^2 \, (\text{psf})$$

For the purpose of determining the diaphragm force, the net effect of Eq. 87 is that the upper half (i.e., only the upper 5 ft) of both perpendicular walls is used to calculate the wall weight. (See Secs. 97 and 116.) The remaining seismic force passes directly into the foundation without being carried by the wall-diaphragm connection. The weight of half the perpendicular walls is

$$W_{\perp \text{wall}} = \left(100 \, \frac{\text{lbf}}{\text{ft}^2}\right)(5 \, \text{ft})(2 \, \text{walls})\left(\frac{50 \, \text{ft}}{\text{wall}}\right)$$
$$= 50,000 \, \text{lbf}$$

(a) Equation 62 is for calculating the base shear passing through to the foundation and should not strictly be used to calculate the force on connections between elements in the building. Equation 73 is for parts and portions and must be used. (See Sec. 110.) For diaphragms, Table 15 substitutes a note and reference to UBC Sec. 2337(b)9 for a value of C_p. That UBC section says (1) that the roof diaphragm is to be designed to resist a portion of the floor forces above it, as weighted by the floor weights, and (2) that the diaphragm force must be within $0.35ZIW_p$ and $0.75ZIW_p$. With the values of $Z = 0.4$ and $I = 1.25$, the limits on diaphragm force are $0.175W_p$ and $0.375W_p$.

Inasmuch as this is a simple one-story building with only one diaphragm, all of the inertial load from the accelerating wall and roof masses must be carried by the wall-roof connection, so Eq. 62 with the wall and diaphragm weights is ultimately used. Checking, $0.175 < 0.229 < 0.375$, (ok).

$$F_{\text{diaphragm}} = 0.229W$$
$$= (0.229)(40,000 \, \text{lbf} + 50,000 \, \text{lbf})$$
$$= 20,600 \, \text{lbf}$$

The shear per foot of diaphragm width, b, is given by Eq. 90.

$$v = \frac{F_{\text{diaphragm}}}{2b} = \frac{20,600 \, \text{lbf}}{(2)(40 \, \text{ft})}$$
$$= 258 \, \text{lbf/ft}$$

(b) Table 17 (corresponding to UBC Table 25-J-1) gives the nail spacing directly. The allowable shear is 375 lbf/ft (> 258 lbf/ft) with a nail spacing of 2.5 in.

(c) The distributed seismic force, w, across the face of the diaphragm is

$$w = \frac{F_{\text{diaphragm}}}{L} = \frac{20,600 \, \text{lbf}}{50 \, \text{ft}}$$
$$= 412 \, \text{lbf/ft}$$

From Eq. 92, the chord force, C, is

$$C = \frac{wL^2}{8b} = \frac{\left(412 \, \frac{\text{lbf}}{\text{ft}}\right)(50 \, \text{ft})^2}{(8)(40 \, \text{ft})}$$
$$= 3220 \, \text{lbf}$$

(d) This is an application of Eq. 88. The net effect is to include the inertial force for accelerating the diaphragm mass and half of all the walls. The diaphragm and perpendicular wall weights have already been determined. Half the weight of the parallel walls is

$$W_{\parallel \text{wall}} = \left(100 \, \frac{\text{lbf}}{\text{ft}^2}\right)(5 \, \text{ft})(2 \, \text{walls})\left(40 \, \frac{\text{ft}}{\text{wall}}\right)$$
$$= 40,000 \, \text{lbf}$$

The total weight is

$$W_{\text{total}} = W_{\text{diaphragm}} + W_{\perp \text{wall}} + W_{\parallel \text{wall}}$$
$$= 40,000 \, \text{lbf} + 50,000 \, \text{lbf} + 40,000 \, \text{lbf}$$
$$= 130,000 \, \text{lbf}$$

The seismic force is

$$V = 0.229W = (0.229)(130{,}000 \text{ lbf})$$
$$= 29{,}770 \text{ lbf}$$

Since the perpendicular walls have no rigidity, all of the seismic force is resisted by the two parallel walls. The shear stress is

$$v = \frac{V}{A} = \frac{29{,}770 \text{ lbf}}{(2 \text{ walls})(8 \text{ in})\left(40 \dfrac{\text{ft}}{\text{wall}}\right)\left(12 \dfrac{\text{in}}{\text{ft}}\right)}$$
$$= 3.88 \text{ psi}$$

(e) Part (d) was an analysis problem, so it was not necessary to include the 150% term required by UBC Sec. 2407(h)4F(i). (See Sec. 117.) However, in determining adequacy, the increase must be included. At 150% of the seismic load, the shear stress would be

$$v_{\text{design}} = (1.5)(3.88 \text{ psi}) = 5.82 \text{ psi}$$

Since this is less than 35 psi, the limit for in-plane shear reinforcement (see Ftn. 114), the wall thickness is adequate.

(f) Since the shear load along the wall was calculated in part (a) to be 258 lbf/ft, the connection between the 10-ft stub wall and the collector would have to carry 30 ft of compressive or tensile loading.

$$C = (30 \text{ ft})\left(258 \dfrac{\text{lbf}}{\text{ft}}\right) = 7740 \text{ lbf}$$

Notice that eliminating 30 ft of parallel wall reduces the accelerating mass but does not reduce the diaphragm force. Only the perpendicular walls affect the diaphragm force.

8

GENERAL STRUCTURAL DESIGN

134 DISTRIBUTING STORY SHEARS TO MEMBERS OF UNKNOWN SIZE

Sections 118 and 120 dealt with distributing seismic forces in frames whose resisting elements (fixed columns, shear walls, etc.) were already designed. (The seismic force to be resisted is the sum of story shears for a particular level and all levels above.) If the cross-sectional area and moment of inertia of these resisting elements are not known (as they will not be initially), assumptions must be made about the amount of lateral force each element carries.

There are several approximate methods (e.g., the portal, cantilever, and Spurr methods discussed in the next sections) that distribute the lateral forces to the resisting elements. These methods eliminate the need to use indeterminate solution methods such as moment distribution.[122]

135 PORTAL METHOD

The *portal method* is ideal for cases where the framing system is regularly spaced. It assumes that all interior columns carry the same shear, while exterior columns carry half the shear of interior columns.[123] Inflection points are assumed to occur at mid-span in each girder

and column. Changes in length due to compression, tension, and deflection are disregarded. Each bay is treated independently of adjacent bays.

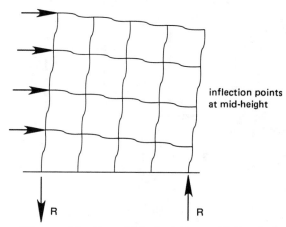

Figure 50 Portal Method Frame Deflection

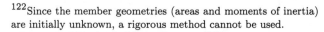

Example 18

The sum of story shears for a particular layer and those above is 110 kips. Use the portal method to find the shears in members 2-3 (i.e., the vertical shear on the horizontal members) and 2-7 (i.e., the horizontal shear on the vertical members).

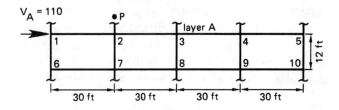

---

[122]Since the member geometries (areas and moments of inertia) are initially unknown, a rigorous method cannot be used.

[123]If the bay sizes (i.e., the distances between columns) are not the same, the shear may be distributed to the columns in proportion to the bay sizes.

Solution

Exterior columns 1-6 and 5-10 carry only half the loads of the interior columns. Therefore, there are a total of four full-strength columns. The horizontal shear carried by members 2-7 and all other interior columns is

$$V_{\text{interior}} = \frac{110 \text{ kips}}{4} = 27.5 \text{ kips}$$

Column 2-7 carries no axial load.

Forces are assumed to be applied at the inflection points of the columns and girders. The vertical girder shears, V_{girder}, are obtained by taking a free-body diagram of joint 2 and summing moments about point 2.

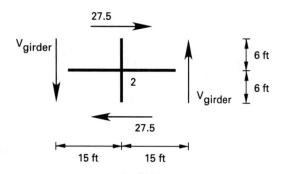

$$\sum M_2 = (2)(27.5 \text{ kips})(6 \text{ ft}) - (2)(V_{\text{girder}})(15 \text{ ft}) = 0$$
$$V_{\text{girder}} = 11 \text{ kips}$$

◇   ◇   ◇   ◇   ◇   ◇   ◇

136 CANTILEVER METHOD

Unlike the portal method, which treats each bay independently of the others, the *cantilever method* assumes the entire floor works together as a unit. Although analysis of the cantilever method must begin at the roof level and work down, this method is preferred for buildings with more than 25 stories. It assumes that the floors remain plane (though not horizontal) and the force in a column is proportional to the distance of the column from the frame's center of gravity. As with the portal method, inflection points of columns and girders are assumed to occur at the mid-lengths.

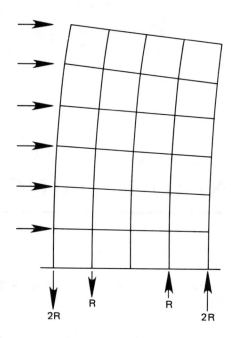

Figure 51   Cantilever Method Frame Deflection

Example 19

The cumulative lateral force at the roof level of the frame shown is 110 kips, as it was in Ex. 18. Use the cantilever method to determine the shear in members 2-3 and 2-7.

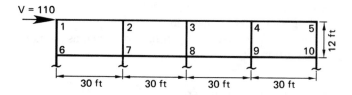

Solution

The center of gravity of this level is located at point Q. Columns 1-6 and 2-7 are in tension. Columns 4-9 and 5-10 are in compression. Column 3-8 is along the neutral axis and is not stressed axially.

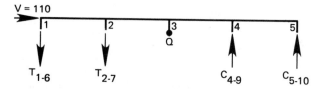

The sum of moments in a member at an inflection point is zero. Summing moments about point Q (halfway between points 3 and 8),

$$\sum M_Q = (6 \text{ ft})(110 \text{ kips}) - (60 \text{ ft})(T_{1\text{-}6})$$
$$- (30 \text{ ft})(T_{2\text{-}7}) - (30 \text{ ft})(C_{4\text{-}9})$$
$$- (60 \text{ ft})(C_{5\text{-}10})$$
$$= 0$$

Since the column forces are proportional to the distance from column 3-8, and all columns are separated by the same distance, it is apparent that the relationships between the tension, T, and compressive, C, forces are

$$T_{1\text{-}6} = 2T_{2\text{-}7} = 2C_{4\text{-}9} = C_{5\text{-}10}$$

Making these substitutions,

$$T_{1\text{-}6} = C_{5\text{-}10} = 4.4 \text{ kips}$$
$$T_{2\text{-}7} = C_{4\text{-}9} = 2.2 \text{ kips}$$

The vertical shear in member 1-2 is equal to the sum of vertical loads to the left of its inflection point.

$$V_{1\text{-}2} = T_{1\text{-}6} = 4.4 \text{ kips}$$

Taking the free body of the inflection points and summing moments about point 1 gives the horizontal shear in member 1-6.

$$\sum M_1 = (6 \text{ ft})(V_{1\text{-}6}) - (15 \text{ ft})(4.4 \text{ kips}) = 0$$
$$V_{1\text{-}6} = 11 \text{ kips}$$

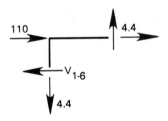

The vertical shear in member 2-3 is

$$V_{2\text{-}3} = 2.2 \text{ kips} + 4.4 \text{ kips} = 6.6 \text{ kips}$$

(This is approximately half the value calculated with the portal method in Ex. 18.)

Summing moments about point 2 gives the horizontal shear in member 2-7.

$$\sum M_2 = (6 \text{ ft})(V_{2\text{-}7}) - (15 \text{ ft})(4.4 \text{ kips})$$
$$- (15 \text{ ft})(6.6 \text{ kips}) = 0$$
$$V_{2\text{-}7} = 27.5 \text{ kips}$$

(This is the same as the value calculated with the portal method in Ex. 18.)

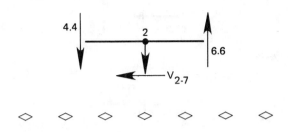

137 SPURR METHOD

Although the cantilever method assumes that the floor will remain plane, it does nothing to ensure a plane floor. Irregular lengthening and shortening of the columns produce secondary bending stresses that can become excessive in tall structures and disturb the planar nature of the floor.

If the height-to-bay length ratio is less than 4, the floor will indeed remain fairly plane. However, when the height-to-bay length ratio exceeds 4, interior column elongation and shortening will cause noticeable floor deflection.

The *Spurr method* uses the cantilever method to calculate girder shears, but girders are given specific strengths in order to justify the assumption that inflection points are located at mid-span.[124] Essentially, the girder moments of inertia are made proportional to the girder shear times the square of the span length.

138 FRAME MEMBER SIZING

Once the column and girder shears and axial loads are known, these members can be sized by traditional methods. For example, a girder can be considered to be loaded at its mid-points (between two sets of columns) by the vertical girder shear acting in opposite directions.

---

[124]This method is named for H. V. Spurr, who published the method in his book *Wind Bracing: The Importance of Rigidity in High Towers*, New York: McGraw-Hill, 1930.

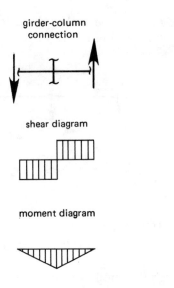

Figure 52   Girder Shear and Moment Diagrams

Another method for sizing frame members is to select the columns and girders so that the drift is limited to the UBC maximum [Sec. 2334(h)2] under the action of the horizontal forces. This method, however, requires making assumptions about the joint rigidity. For example, assuming that a bent consisting of a girder of length L and two columns of height h must limit the drift to 0.5% of the story height when a total shear force of V is experienced, and the joints are all rigid, the drift criterion is

$$\frac{Vh^2}{12E}\left(\frac{h}{I_{\text{column}}} + \frac{L}{I_{\text{girder}}}\right) = 0.005h \qquad [98]$$

There are many possible combinations of column and girder strengths that will satisfy this criterion.

9
DETAILS OF SEISMIC-RESISTANT CONCRETE STRUCTURES

139 CONCRETE CONSTRUCTION DETAILS

Concrete itself has poor ductility in shear and is therefore a brittle material. This limited ductility, combined with its higher mass compared to steel (higher mass increases the seismic force) and lower tolerance to errors in design and workmanship, usually makes concrete a second choice for high-rise construction. However, concrete ductile moment-resisting frames now make it possible to design structures with the ductility and energy dissipation capability of steel structures. Moreover, a concrete building may be up to 20% less expensive to construct.[125] The primary disadvantage of concrete is its weight, which increases the seismic forces. Some advantage can be gained, however, by using lightweight concrete.

Concrete structures can be constructed to behave in a ductile manner. Such structures are loosely referred to as having been constructed of "ductile concrete," "special concrete," or "California concrete." In order to make concrete into a ductile structure, much attention must be given to connection and confinement details.[126] "Ordinary concrete," or concrete construction that does not behave in a ductile way, cannot be used in California.

The two most important concepts in designing ductile concrete are (1) continuity and (2) confinement. Concrete members must not pull apart, and they must not disintegrate when the core becomes cracked or crushed.

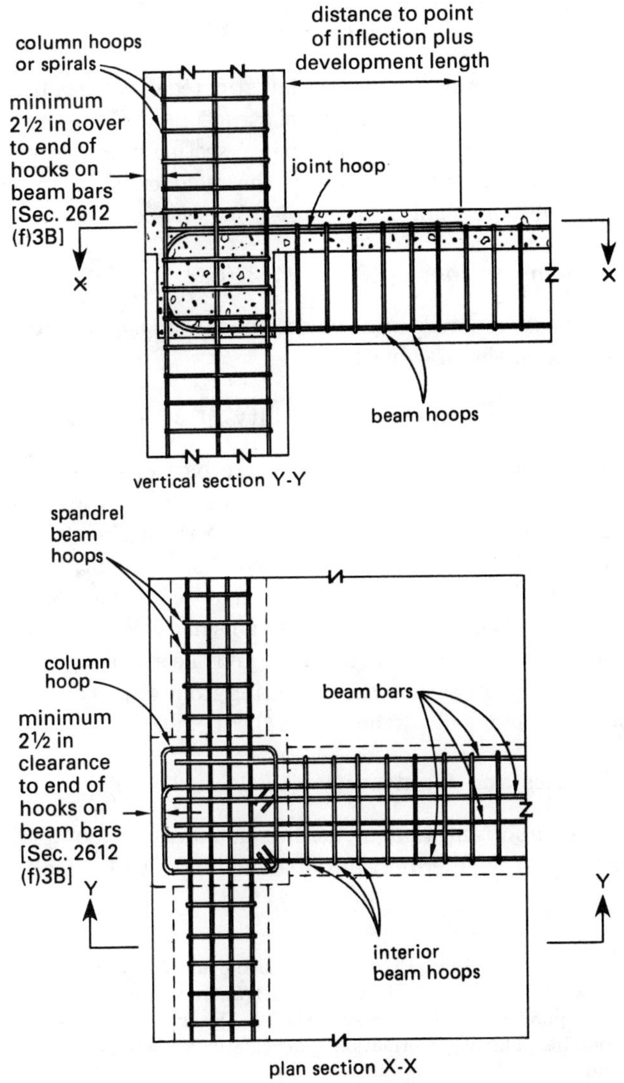

---

[125]The assumed economic superiority of steel over concrete is traditional and is a very controversial issue; it is not, by any means, an absolute fact.

[126]It is not always easy, however, to get all the special steel reinforcing and confinement into the beam-column connection area. The congestion, called a "tangle" by some, can be considerable.

vertical section Y-Y

plan section X-X

Figure 53 Typical Ductile Frame Joint for Region of High Seismic Risk [UBC Sec. 2625(g)]

The UBC discusses the required details of ductile concrete in its Secs. 2337 and 2625.[127] The following lettered items list some of the more important provisions. Because all provisions have been greatly simplified in this book, the UBC references should be used for clarification. In almost all cases, other provisions also apply. (In some of the following cases, the accompanying figures contain additional information about special seismic provisions.)

The special provisions in Sec. 2625 of the UBC apply only to frames (including beams, columns, and slabs) that resist earthquake loading. Shear walls and all other features designed according to regular sections of UBC Chapter 26 are considered adequate. UBC Sec. 2625(i) still requires members that are not part of the lateral force-resisting system to satisfy minimum reinforcement requirements.

A. Orthogonal Effects   [UBC Sec. 2337(a)]

Orthogonal effects should be investigated in cases of torsional irregularity, nonparallel structural systems, and where a member is part of two intersecting lateral force-resisting systems (e.g., where the member is a corner column more than lightly loaded by seismic forces).[128]

B. Connections   [UBC Sec. 2337(b)3]

Connection details must be designed by an engineer and shown on the drawing.

C. Deformation Compatibility   [UBC Sec. 2337(b)4]

Not all members in a structure are part of the lateral force-resisting system. However, members that are not part of that system must be able to resist the strains induced by the $3R_w/8$ magnified displacement. Unless special transverse reinforcement (e.g., hoops and crossties) is used, compressive strains in a concrete member must not exceed 0.003 in/in, and the shear strength must be greater than that resulting from displacement-induced moments atthe end of the member.

D. Ties and Continuity   [UBC Sec. 2337(b)5]

It is important that structural members do not pull apart. All smaller portions of a building must be tied to the rest of the building so that a force of $Z/3$ times the weight of the smaller portion is resisted. Connections of beams, girders, and trusses must resist a horizontal force *parallel* to the member not less than $Z/5$ times the gravity load reaction from dead and live loads.

E. Collector Elements   [UBC Sec. 2337(b)6]

Concrete collector elements must rely on reinforcing steel to carry drag forces into shear walls.

F. Concrete Frames   [UBC Sec. 2337(b)7]

Concrete frames used in lateral force-resisting systems in zones 3 and 4 must be special moment-resisting frames (SMRF).

G. Anchorage to Concrete and Masonry Walls   [UBC Sec. 2337(b)8]

Connections to and between walls must be capable of resisting the greater of 200 pounds per foot of wall or the force specified in UBC Sec. 2336. When anchor spacing exceeds 4 ft, walls must resist bending between the anchors.

H. Boundary Elements of Shear Walls and Diaphragms   [UBC Sec. 2625(f)]

The reinforcement ratio cannot be less than 0.0025 along both the longitudinal and transverse axes of shear walls and diaphragms. Reinforcement spacing cannot exceed 18 in. (See Fig. 54.)

Boundary elements of shear walls must not fail due to deformations caused by a major earthquake, particularly when the design is governed by flexure. The design should specify special detailing to prevent tensile fracture as well as compressive crushing and buckling. This is accomplished by requiring two curtains of shear reinforcement for walls with factored shear stress in excess of $2\sqrt{f_c'}$. Additional reinforcement is required at the edges of all shear walls and diaphragms, as well as at the boundaries of all openings.

I. Seismic Hooks, Crossties, and Hoops   [UBC Sec. 2625(b)]

The details of *seismic hooks, crossties,* and *hoops* are shown in Fig. 55. For a seismic hook, the bend must be at least 135 degrees with an extension past the bend of at least 6 bar diameters (but not less than 3 in).

J. Strength Reduction Factor   [UBC Sec. 2625(c)3]

In general, the normal strength reduction factor, ϕ, defined in UBC Sec. 2609(d) is applicable. The strength reduction factor is 0.85 for cast-in-place concrete diaphragms and beam-column joints. The strength

---

[127]The material in UBC Sec. 2625 is equivalent to the material contained in Chapter 21 of the ACI (American Concrete Institute) publication 318. However, the Blue Book commentary also contains valuable information, particularly for seismic zones 3 and 4.

[128]The column is more than lightly loaded if its factored axial load due to seismic forces is greater than 20% of the column axial strength.

reduction factor is 0.50 for frame members with $P_u > 0.10 f'_c A_g$ and transverse reinforcement not meeting UBC Sec. 2625(e)4.

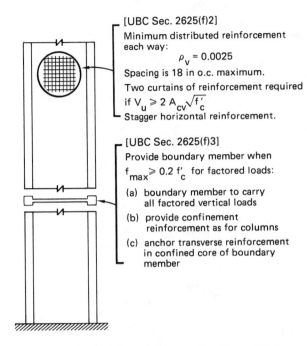

[UBC Sec. 2625(f)2]

Minimum distributed reinforcement each way:
$$\rho_v = 0.0025$$
Spacing is 18 in o.c. maximum.

Two curtains of reinforcement required if $V_u \geq 2 A_{cv} \sqrt{f'_c}$

Stagger horizontal reinforcement.

[UBC Sec. 2625(f)3]

Provide boundary member when $f_{max} \geq 0.2 f'_c$ for factored loads:

(a) boundary member to carry all factored vertical loads

(b) provide confinement reinforcement as for columns

(c) anchor transverse reinforcement in confined core of boundary member

Figure 54   Requirements for Shear Walls and Boundary Members

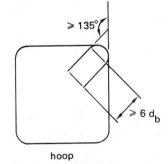

seismic hook

$\geq 135°$

$\geq 6\,d_b$

$90°$

$\geq 6\,d_b$

crosstie

$\geq 135°$

$\geq 6\,d_b$

hoop

Figure 55   Seismic Hook, Crosstie, and Hoop [UBC Sec. 2625(b)]

K. Combined Loading Load Factors
[UBC Sec. 2625(c)4]

The ultimate strength for members with different combinations of loading are

$$U = 1.4D + 1.7L \qquad [99]$$

$$U = 1.4(D + L + E) \qquad [100]$$

$$U = 0.9D \pm 1.4E \qquad [101]$$

L. Concrete Strength   [UBC Secs. 2625(c)5A and 5B]

The compressive strength, f'_c, of normal-weight concrete cannot be less than 3000 psi. The compressive strength of lightweight concrete cannot exceed 4000 psi unless experimental data are submitted to show that it has properties equivalent to normal-weight concrete. However, the compressive strength of lightweight concrete used in calculations cannot exceed 6000 psi under any circumstances.

UBC Sec. 2625(j) requires that concrete used in moment frames be inspected by a qualified inspector.

M. Steel Reinforcement   [UBC Sec. 2625(c)6]

All reinforcing steel must be of the deformed variety; plain bars are not permitted. Steel used must be low-alloy complying with ASTM A706. ASTM A615 grades 40 or 60 may be used if (1) the actual yield strength is no more than 18,000 psi higher than the specified yield, and (2) the actual ultimate tensile stress is at least 125% of the actual yield strength. Restrictions on welding apply.

N. Frame Flexural Members   [UBC Sec. 2625(d)]

(This UBC section is specifically for flexural members used in lateral force-resisting systems.) The UBC imposes a geometric restraint (i.e., a width/depth ratio greater than or equal to 0.3) and other constraints [Secs. 2625(d)1C and 2625(d)1D] in order to reduce lateral instability and ensure a transfer of moment during inelastic response. (See Fig. 56.)

Seismic hoops (confinement) are required in all areas of flexural members where yielding is expected (e.g., at the ends of built-in beams and at other plastic hinge points). The first hoop must be within 2 in of the face of the supporting member. Maximum spacing must be less than (a) $d/4$, (b) 8 times the diameter of the smallest longitudinal bar, (c) 24 times the hoop bar diameter, and (d) 12 in. Where hoops are not required, stirrup spacing with seismic hooks must be spaced at no more than $d/2$ throughout the member [Sec. 2625(d)3].

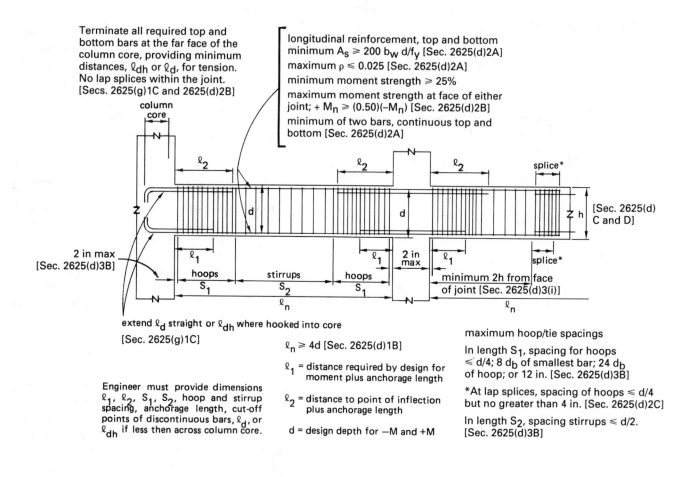

Figure 56   Special Flexural Member Reinforcement
[UBC Sec. 2625(d)]

O. Top and Bottom Reinforcement
[UBC Secs. 2625(d)2A and 2B]

At least two top and two bottom bars must run the entire length of the member. For both top and bottom steel, the amount of reinforcement steel must be greater than $(200/f_y)b_w d$, and the maximum steel is $0.025b_w d$. At the face of a joint, the positive moment strength cannot be less than 50% of the negative moment strength. At any section along the member length, neither the positive nor the negative moment strengths can be less than 25% of the corresponding strength at a joint.

P. Tension Lap Splices   [UBC Sec. 2625(d)2C]

Lap splices are classified as Class A (with a lap length of one development length) and Class B (with a lap length of 1.3 development lengths) [UBC Sec. 2612(p)]. (The Class C splice (1.7 development lengths) was eliminated in 1989, and (in some cases) the basic development length was increased.)

Tension lap splices are prohibited in flexural members (1) within beam-column joints, (2) within a distance from the face of a joint equal to twice the member depth, and (3) where flexural yielding is anticipated.

Lap splices are not permitted where flexural yielding is expected (i.e., at plastic hinge points) because such splices are not reliable when the loading repeatedly exceeds the yield strength of the reinforcement. Where permitted, they are proportioned as tension lap splices. Class B splices must be used unless double (or more) of the required reinforcement steel is provided [UBC Sec. 2612(b)2].

Q. Welded Splices and Mechanical Connections
[UBC Secs. 2625(d)2D and 2625(e)3B]

Welded splices and approved mechanical connections can be used and are required for longitudinal No. 14 and 18 bars. Such splices are permitted at any other section provided they (1) are on alternate bars in each

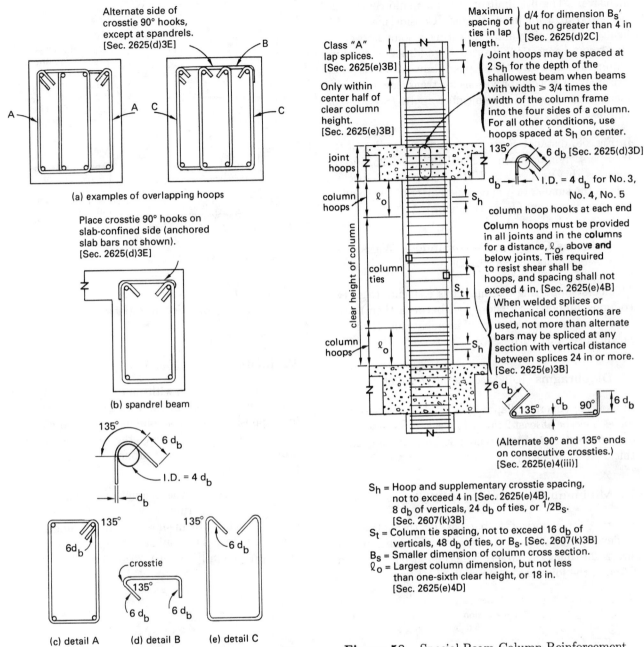

Alternate side of crosstie 90° hooks, except at spandrels. [Sec. 2625(d)3E]

(a) examples of overlapping hoops

Place crosstie 90° hooks on slab-confined side (anchored slab bars not shown). [Sec. 2625(d)3E]

(b) spandrel beam

135°   6 d_b

I.D. = 4 d_b

d_b

135°   135°

6d_b 6 d_b

crosstie

135°

6 d_b 6 d_b

(c) detail A   (d) detail B   (e) detail C

Stirrups required to resist shear shall be hoops. Throughout the length of flexural members where hoops are not required, stirrups shall be spaced at no more than d/2 and shall have seismic hooks.

Figure 57   Details of Transverse Confinement Reinforcement in Flexural Members [UBC Sec. 2625(b)]

Maximum spacing of ties in lap length.   d/4 for dimension B_s' but no greater than 4 in [Sec. 2625(d)2C]

Class "A" lap splices. [Sec. 2625(e)3B]

Only within center half of clear column height. [Sec. 2625(e)3B]

Joint hoops may be spaced at 2 S_h for the depth of the shallowest beam when beams with width $\geq$ 3/4 times the width of the column frame into the four sides of a column. For all other conditions, use hoops spaced at S_h on center.

joint hoops

column hoops

column ties

clear height of column

column hoops

135°   6 d_b [Sec. 2625(d)3D]

d_b I.D. = 4 d_b for No. 3, No. 4, No. 5 column hoop hooks at each end

Column hoops must be provided in all joints and in the **columns** for a distance, ℓ_o, above **and** below joints. Ties required to resist shear shall be hoops, and spacing shall not exceed 4 in. [Sec. 2625(e)4B]

When welded splices or mechanical connections are used, not more than alternate bars may be spliced at any section with vertical distance between splices 24 in or more. [Sec. 2625(e)3B]

ℓ_o S_h

S_t

ℓ_o S_h

6 d_b

135° d_b 90°   6 d_b

(Alternate 90° and 135° ends on consecutive crossties.) [Sec. 2625(e)4(iii)]

S_h = Hoop and supplementary crosstie spacing, not to exceed 4 in [Sec. 2625(e)4B], 8 d_b of verticals, 24 d_b of ties, or $^1/2B_s$. [Sec. 2607(k)3B]

S_t = Column tie spacing, not to exceed 16 d_b of verticals, 48 d_b of ties, or B_s. [Sec. 2607(k)3B]

B_s = Smaller dimension of column cross section.

ℓ_o = Largest column dimension, but not less than one-sixth clear height, or 18 in. [Sec. 2625(e)4D]

Figure 58   Special Beam-Column Reinforcement [UBC Sec. 2625(e)]

R. Members with Bending and Axial Loads
[UBC Sec. 2625(e)]

The minimum member dimension is 12 in. The ratio of the shortest-to-longest dimension cannot be less than 0.4. Class A tension lap splices are allowed within the center half of the member's length [Sec. 2625(e)3B]. The total column moment-resisting strength at a joint must be greater than 6/5 of the total girder moment-resisting strength at a joint [Sec. 2625(e)2]. The longitudinal reinforcement ratio must be between 0.01 and 0.06,

layer and (2) have 24 in minimum distance between two adjacent splices. Welding of designed reinforcement for any purpose other than approved splicing is prohibited. Welding of stirrups, ties, inserts, and other elements to the longitudinal bars is prohibited [Sec. 2625(c)6].

inclusive. (The 6% limit on column reinforcement ratio is too high for most designs. Considering the dif-
ficulty in designing and fabricating a joint as well as
the "joint congestion" caused by additional ductility
requirements, a practical upper limit of 4% is more
reasonable.) Special provisions for transverse reinforcement (confinement) in the member apply. (See Figs. 58
and 59.)

In Fig. 58, distance S_h is the maximum spacing of the
transverse reinforcement. The UBC [Sec. 2625(e)4B]
specifies this as 4 in. ACI 318, Sec. 21.4.4.2 specifies
this as the smaller of 4 in or one-quarter of the smallest
member dimension (B_s in the figure).

S. Columns Carrying Discontinuous Walls
[UBC Secs. 2625(e)4E and 2625(f)6]

Members supporting discontinuous elements (e.g., two
columns supporting a short wall between them) must
have special transverse reinforcement for their full
height. (See Fig. 60.)

T. Diaphragms   [UBC Secs. 2625(f)7 and 8]

Monolithic concrete diaphragms used to resist seismic
forces must be at least 2 in thick. Cast-in-place toppings
over precast elements should not be less than 2.5 in
thick.

U. Minimum Column Size   [UBC Sec. 2625(g)1D]

In order for a horizontal bar passing through a joint
to develop its full bond strength, the minimum column
size is 20 times the largest longitudinal bar extending
through the joint in the direction of loading.

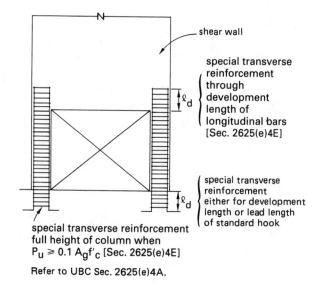

special transverse reinforcement
full height of column when
$P_u \geq 0.1\,A_g f'_c$ [Sec. 2625(e)4E]

Refer to UBC Sec. 2625(e)4A.

Figure 60   Columns Carrying a Discontinuous Wall
[UBC Sec. 2625(f)6]

V. Joints   [UBC Sec. 2625(g)]

Joints must be able to withstand forces of $1.25 f_y$ in the
longitudinal reinforcement. Flexural members framing
into opposite sides of a column must have continuous
reinforcement to pass through the column joint. Special
provisions for confinement around the joint apply. See
Fig. 53.

A 50% reduction in the amount of reinforcement confining a joint is permitted when all four sides of the
joint are confined by members (beams) framing into the
joint (column). The confinement spacing may also be
increased to 6 in [UBC Sec. 2625(g)2B].

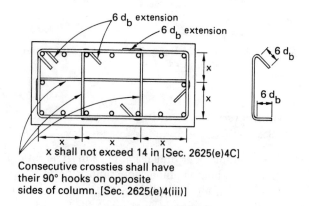

x shall not exceed 14 in [Sec. 2625(e)4C]
Consecutive crossties shall have
their 90° hooks on opposite
sides of column. [Sec. 2625(e)4(iii)]

Figure 59   Details of Transverse Confinement
Reinforcement in Beam-Columns
[UBC Sec. 2625(e)]

10

DETAILS OF SEISMIC-RESISTANT STEEL STRUCTURES

140 STEEL CONSTRUCTION DETAILS

The most common structural system for high-rise buildings in seismically active areas is the ductile steel frame. Ductile frame behavior is much easier to achieve with steel construction than with concrete construction because steel is intrinsically a ductile material. Design emphasis is therefore shifted away from ensuring that the material itself will behave in a ductile manner to ensuring that the structural frame will operate properly.

Ductile frame operation is accomplished by extensive use of moment-resisting connections known as *type I connections*.[129] These connections transmit column moments to beams and girders, forcing those members to carry the moments. It is easier for beams and girders to resist moments because their lengths (and, hence, their moment arms) are long. Without moment-resisting joints, the columns would most likely fail (by web crippling, for example) before the beams and girders became fully stressed. Therefore, a beam-column connection must be able to transmit without failure a moment equal to the plastic capacity of the beam. Frames that use joints that do this are known as *special moment-resisting frames (SMRF)*.[130] (In Fig. 61, the abbreviations FP and T&B refer to full-penetration welds at the tops and bottoms of the beam flanges.)

---

[129] *Type II joints* do not develop any appreciable moments. Steel structures in any seismic zone can use type II joints. (See Sec. 96.) This is typical of braced-frame designs.

[130]The term *ductile moment-resisting space frame (DMRSF)*, once used extensively, is no longer commonly used.

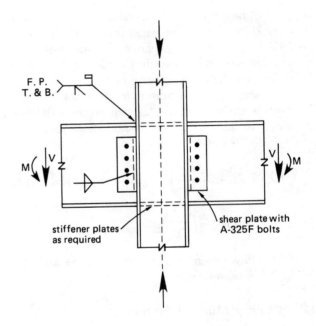

Figure 61   Typical Moment-Resisting Joint

A. Steel Properties   [UBC Sec. 2710(d)1]

With limited exceptions, only certain steels (i.e., A36, A441, A500, A501, grades 42 and 50 of A572, and A588) are permitted in seismic force-resisting frames. Structural steel conforming to A283 (grade D) can be used for base plates and anchor bolts. Other steels not listed and those with yield strengths in excess of 50 ksi cannot be used.

B. Member Strengths   [UBC Sec. 2710(d)2]

The term *strength* (as in "full strength") means that the members must be capable of developing the following forces or moments:

◇ moment in flexural members　$M = ZF_y$
◇ shear in flexural members　$V = 0.55F_y dt$
◇ axial compression in
　flexural members　$P = 1.7F_a A$
◇ axial tension in
　flexural members　$P = F_a A$
◇ bolts　1.7 × allowable
◇ full-penetration welds　$F_y A$
◇ partial-penetration welds　1.7 × allowable
◇ fillet welds　1.7 × allowable

C. Column Requirements   [UBC Sec. 2710(e)]

Loads must be supported at allowable stress limits, with the one-third stress increase permitted by UBC Sec. 2303(d). Since the integrity of a structural steel system that has experienced ductile yielding is strongly dependent on the axial capacity of the columns, with limited exceptions, columns in seismic zones 3 and 4 must have the strength (as defined in UBC Sec. 2710(d)2) to support the load combinations given.[131]

$$1.0D + 0.7L + 3\left(\frac{R_w}{8}\right)E \quad \text{[axial compression]} \quad [102]$$

$$0.85D + 3\left(\frac{R_w}{8}\right)E \quad \text{[axial tension]} \quad [103]$$

Other specifications relating to column splices and slenderness must also be met.

D. Ordinary Moment Frames   [UBC Sec. 2710(f)]

Ordinary moment frames (OMFs) are permitted, but the Blue Book commentary discourages such use. It is likely, in any case, that OMFs will be heavier and thus costlier than SMRFs, since OMFs resist seismic forces elastically. OMFs may have applications where seismic forces are low and the design is controlled by wind.

E. Special Moment-Resisting Frames
[UBC Sec. 2710(g)]

This UBC section is designed to ensure that joints and members behave in a ductile manner in SMRFs. The areas covered are the beams and columns in which plastic hinges can form and in the joint *panel zones* in which

shear yielding can occur. With specific limitations, inelastic behavior in these areas is permitted. However, the joint itself cannot yield [Sec. 2710(g)2].

[Sec. 2710(g)2]

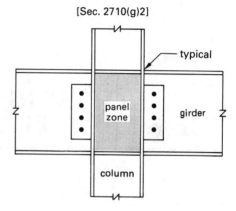

Figure 62　Panel Zone

The beam-column connection must have minimum strength (i.e., strength equal to the beam in flexure). The code says this strength is achieved when beam flanges are full-penetration welded to columns and the beam web-to-column portion of the connection alone is able to support the gravity and seismic shear force. Bolted connections and other methods are also permitted if justified.[132]

Doubler plates can be used, with restrictions, to reduce panel zone shear stress or web depth/thickness ratio. *Continuity plates* must satisfy a certain width/thickness ratio [Secs. 2710(g)2C and 2710(g)4].[133]

A minimum *strength ratio* is given in order to prevent column failures at joints when the beams are "stronger" than the columns. This is the "strong column-weak beam" test. Some exceptions are permitted, however [Sec. 2710(g)5].

Within strength requirements, *trusses* may be used as horizontal members in SMRFs [Sec. 2710(g)6].

In order to withstand reverse bending moments, intersecting beams or bottom flange diagonal bracing are required.

F. Girder-Column Joint Restraints
[UBC Sec. 2710(g)7]

Restrictions are given to ensure that moment frames are capable of reaching their seismic design capacity.

---

[131]Despite appearances, this is not ultimate strength (plastic or limit) design.

[132]Full-penetration welding of beam flanges to columns is expensive and is not the standard method preferred by erectors.

[133]The Blue Book commentary strongly suggests the use of continuity plates.

Specifically, if the columns remain elastic, column flanges require lateral support only at the level of the top girder flange. (The UBC describes the conditions under which the column can be assumed to remain elastic.) If the column does not remain elastic, column flange support is required at the tops and bottoms of the girder flanges.

G. Braced Frames   [UBC Sec. 2710(h)]

Various requirements, including slenderness ratios, are given for braced frames. Different systems of braced frames are used: *diagonal bracing* (where diagonals connect the joints in adjacent levels); *chevron bracing* (where a pair of braces terminate at a single point within the clear beam span); *V bracing* (a form of chevron bracing that intersects a beam from above); *inverted V bracing* (a form of chevron bracing that intersects a beam from below); *K bracing* (where a pair of braces located on one side of a column terminate at a single point within the clear column height); and, *X bracing* (where a pair of diagonal braces cross near mid-length of the bracing members). Chevron bracing can be used when certain conditions are met [Sec. 2710(h)4A]. With limited exceptions, K bracing is not permitted in seismic zones 3 and 4 [Sec. 2710(h)4B], but the lateral seismic load must be increased by multiplying by a factor of 1.5 when designing bracing members [Sec. 2710(h)4A(i)].

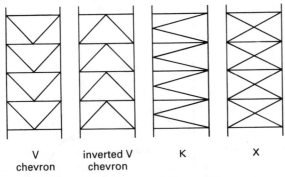

| V | inverted V | | |
| chevron | chevron | K | X |

Figure 63   Types of Braced Frames

141 ECCENTRICALLY-BRACED FRAMES
[UBC Sec. 2710(i)]

Figure 64 illustrates the basic types of *eccentric braced frames (EBF)* in which at least one end of each brace intersects a beam at a point away from the column girder joint.[134] The short section of girder between the brace

---

end and the column is known as the *link*. Links designed to yield are known as *link beams*.[135] Types (a) and (b) are known as *end link EBFs*, and type (c) is known as the *center link EBF*. EBFs may be inverted, and different EBF systems can be mixed. Elastic performance, rather than plastic performance, is used in the design and analysis of most EBFs, since all members are steel and load resistance factor design (LRFD) is not in widespread use.

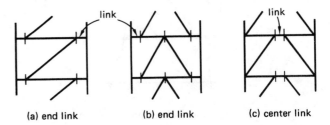

| (a) end link | (b) end link | (c) center link |

Figure 64   Types of Eccentric Braced Frames

The concept behind EBFs is that in low-to-moderate ground shaking, a frame using them performs as a braced frame rather than as a moment frame. Therefore, the structure experiences small drifts, little if any architectural damage, and no structural damage. The link is specifically designed to yield in a major seismic event, thereby absorbing large quantities of seismic energy and preventing buckling of the other bracing members. To limit yielding to the link beam requires attention to detail at the connection. Figure 65 illustrates typical details. (Notice the use of web stiffeners in the links to keep the web from buckling. These may not be needed in every instance.)

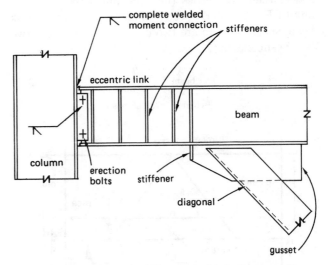

Figure 65   Detail of Eccentric Link Construction

---

[134]This is exactly the type of design that engineers are taught to avoid in traditional classes on structures. However, this is a new area pioneered by the works of Egor P. Popov. One of the earliest California buildings to use this type of feature is the 16-floor Bank of America Regional Office and Branch Bank Building in San Diego, which Popov assisted in designing.

[135]The distinction between a *link* and a *link beam* is not consistently made.

There is also some evidence that the beam-to-column connection should be at the lower end of the diagonal rather than at the top. Tests seem to show that the high ends are always stressed to yielding while the links at the low end remain elastic. This is illustrated in Fig. 66.

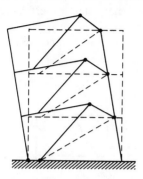

Figure 66 Typical Deflections of Eccentric Links

Section 2710(i) of the UBC covers the design of EBFs. Shear in the link beam and bay drift are the primary design variables. Shear stress in the link beam is limited to $(0.80)(0.55)F_y = 0.44F_y$ [UBC Secs. 2710(d)2 and 2710(i)4]. The braces impose large axial forces in the beams between two brace points. Therefore, critical bending/axial stress loads usually occur outside the link beam.

The shear force on a link beam—either end link or center link—is equal to the story shear times the story height divided by the distance between the column center lines (i.e., the *bay length*). For a given story height and distance between columns, frames with the same story shear will have the same link beam shear, regardless of their geometries (i.e., end link EBF or center link EBF). The load in each column is equal to the link beam shear. (There are two columns to carry the two link beam shears.)

$$V_{\text{link beam}} = \frac{F_x h}{L} \qquad [104]$$

$$P_{\text{column}} = V_{\text{link beam}} \qquad [105]$$

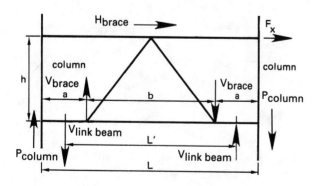

Figure 67 End Link Beam

Figure 67 illustrates an *end link EBF*, also known as a *Type (a) EBF*. The location of the link beam's inflection point initially must be assumed until all member sizes have been determined.[136] A location at mid-length of the link beam is a reasonable initial assumption. The fraction of the story shear carried by the inclined braces compared to that carried by direct moment frame action (approximately 0.75) is equal to the ratio of the distance between the points of inflection on the link beams (shown in Fig. 67 as distance L') to the bay length, b.

$$V_{\text{brace}} = \frac{V_{\text{link beam}}L'}{b} \qquad [106]$$

$$H_{\text{brace}} = \frac{F_x L'}{L} \qquad [107]$$

For a *center link beam*, also known as a *Type (b) EBF*, the beam-to-column connection is normally "pinned" in the frame's plane, and all of the lateral shear is taken by the braces. The link beam point of inflection is at the (vertical) centerline of the bay.[137]

$$V_{\text{link beam}} = \frac{F_x h}{L}$$

$$V_{\text{brace}} = \frac{V_{\text{link beam}}L}{2a} \qquad [108]$$

$$H_{\text{brace}} = \frac{F_x}{2} \qquad [109]$$

$$P_{\text{column}} = V_{\text{link beam}}$$

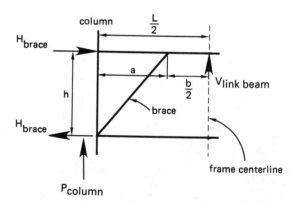

Figure 68 Center Link Beam

---

[136]This is a direct result of the need to weld the beam-column connection in order to obtain a moment-resisting condition.

[137]This is a direct result of the fact that the beam-column connection can be pinned.

11

DETAILS OF SEISMIC-RESISTANT MASONRY STRUCTURES

142 MASONRY CONSTRUCTION DETAILS

Masonry design continues to be based on allowable stress criteria and is covered in UBC Chapter 24. Special provisions for seismic resistance are given in UBC Sec. 2407(h). In particular, UBC Sec. 2407(h)4 is concerned with provisions for seismic zones 3 and 4.

The strength of masonry structures is particularly sensitive to construction methods and quality. Masonry that is constructed under the watchful eye of an expert or other qualified person (as defined in UBC Sec. 306(b)) is entitled to higher design stresses. In particular, the masonry design stresses listed in UBC Sec. 2406(c)2 must be reduced by 50% if *special inspection* is not provided [UBC Sec. 2406(c)1].

The following important points must be observed in masonry design for structures in seismic zones 3 and 4.

1. Masonry shear walls can be designed using ultimate strength theory [UBC Sec. 2412(a)].

2. Lateral loads must be increased by a factor of 1.5 when using the allowable stress method to design masonry shear walls [UBC Sec. 2407(h)4F(i)].

3. Types O and N mortar are not permitted [UBC Secs. 2407(h)3A and 2407(h)4A].

4. Unless higher values are determined from testing, the maximum compressive strengths, f'_m, are 1500 psi for concrete masonry and 2600 psi for clay masonry [UBC Sec. 2407(h)4].

5. #4 bars are the smallest that may be used for vertical reinforcement at each corner or wall intersection and at the edge of any openings. Maximum bar spacing is 4 ft [UBC Sec. 2407(h)3B].

6. #4 bars are the smallest that may be used for horizontal reinforcement at the tops and bottoms of walls and at the tops and bottoms of any openings. Maximum bar spacing is 10 ft [UBC Sec. 2407(h)3B].

7. The minimum steel ratio for vertical and horizontal reinforcement is 0.0007 for stack bond (open-end) concrete masonry units (CMUs) and 0.0015 for closed-cell CMUs [UBC Sec. 2407(h)4D].

12

DETAILS OF SEISMIC-RESISTANT WOOD STRUCTURES

143 WOOD SHEAR WALL AND PLYWOOD DIAPHRAGM DESIGN CRITERIA

Design of wood shear walls and plywood diaphragms requires consideration of diaphragm ratios, horizontal and vertical diaphragm shears, and connector/fastener values.[138] These topics are covered individually in this and the following sections.

Figure 69 illustrates the types of plywood sheathing used to construct diaphragms. *Diagonal sheathing*, consisting of 1-in (nominal) sheathing boards laid at an angle of approximately 45 degrees to the supports, can also be used [UBC Sec. 2513(b)].

In addition, there are special requirements for seismic zones 3 and 4, introduced in the following subsections. Most of these are already common construction practice, but some are state-of-the-art innovations.

A. Framing   [UBC Sec. 2513(e)1B]

Collector members (i.e., drag struts) are required. Openings in diaphragms require perimeter framing. Such perimeter framing must be detailed to distribute shearing forces along its length. Diaphragm plywood cannot be used to splice the perimeter members. Diaphragm chords must be in the plane of the diaphragm unless it can be shown that chords in other locations of the walls will work. (See Fig. 70.)

---

[138]The UBC also specifies in Sec. 2513(a) that diaphragm deflection must be limited to the "permissible deflection" of attached distributing and resisting elements such as walls. *Permissible deflection* is defined as a deflection that maintains the structural integrity when loaded, that is, deflection that can support the loads without endangering the occupants. (See Sec. 129.)

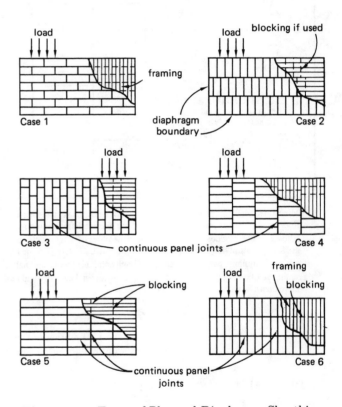

Figure 69   Types of Plywood Diaphragm Sheathing

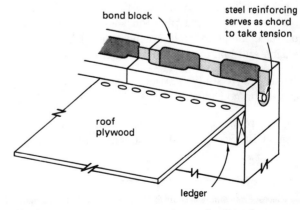

Figure 70   Chord in the Plane of a Diaphragm

Table 17
Allowable Shear for Horizontal Plywood Diaphragms with Douglas Fir-Larch or Southern Pine Framing[1] (pounds per foot)
[UBC Table 25-J-1]

PLYWOOD GRADE	Common Nail Size	Minimum Nominal Penetration in Framing (in inches)	Minimum Nominal Plywood Thickness (in inches)	Minimum Nominal Width of Framing Member (in inches)	BLOCKED DIAPHRAGMS — Nail spacing at diaphragm boundaries (all cases), at continuous panel edges parallel to load (Cases 3 and 4) and at all panel edges (Cases 5 and 6)				UNBLOCKED DIAPHRAGM — Nails spaced 6" max. at supported end	
					6	4	2½/2	2	Load perpendicular to unblocked edges and continuous panel joints (Case 1)	Other configurations (Cases 2, 3 and 4)
					Nail spacing at other plywood panel edges					
					6	6	4	3		
STRUCTURAL I	6d	1¼	5/16	2	185	250	375	420	165	125
				3	210	280	420	475	185	140
	8d	1½	3/8	2	270	360	530	600	240	180
				3	300	400	600	675	265	200
	10d[3]	1⅝	15/32	2	320	425	640	730	285	215
				3	360	480	720	820	320	240
C-D, C-C, STRUCTURAL II and other grades covered in U.B.C. Standard No. 25-9	6d	1¼	5/16	2	170	225	335	380	150	110
				3	190	250	380	430	170	125
			3/8	2	185	250	375	420	165	125
				3	210	280	420	475	185	140
	8d	1½	3/8	2	240	320	480	545	215	160
				3	270	360	540	610	240	180
			15/32	2	270	360	530	600	240	180
				3	300	400	600	675	265	200
	10d[3]	1⅝	15/32	2	290	385	575	655	255	190
				3	325	430	650	735	290	215
			19/32	2	320	425	640	730	285	215
				3	360	480	720	820	320	240

[1] These values are for short-time loads due to wind or earthquake and must be reduced 25 percent for normal loading. Space nails 10 inches on center for floors and 12 inches on center for roofs along intermediate framing members.

   Allowable shear values for nails in framing members of other species set forth in Table No. 25-17-J of U.B.C. Standards shall be calculated for all grades by multiplying the values for nails in STRUCTURAL I by the following factors: Group III, 0.82 and Group IV, 0.65.

[2] Framing at adjoining panel edges shall be 3-inch nominal or wider and nails shall be staggered where nails are spaced 2 inches or 2½ inches on center.

[3] Framing at adjoining panel edges shall be 3-inch nominal or wider and nails shall be staggered where 10d nails having penetration into framing of more than 1⅝ inches are spaced 3 inches or less on center.

B. Plywood   [UBC Sec. 2513(e)1C]

Plywood must be of the exterior glue type. Sheets must be at least 4 ft by 8 ft, except at edges of the diaphragm where pieces with a minimum dimension of 2 ft may be used.[139] Diaphragm plywood sheathing can be used to splice members that are not placed in cross-grain tension and bending from the splicing nails.

Grade 2-M-W (only) particleboard can also be used in diaphragm construction [UBC Sec. 2513(e)1E].

Though plywood can be used for splicing members in shear [UBC Sec. 2513(e)1C], it cannot be used to splice collector and perimeter members that carry tension and compression [UBC Sec. 2513(e)1B].

C. Requirements for Resisting Horizontal Forces in Concrete/Masonry Buildings
[UBC Sec. 2515(b)]

In buildings with concrete or masonry walls, horizontal wood floor and roof trusses and diaphragms can be used to resist seismic and wind forces as long as such forces are not resisted by intended torsion or rotation of the trusses or diaphragms.[140] Vertical shear walls can be used to resist horizontal shear in buildings up to two stories in height as long as certain requirements are met.

---

[139] Pieces even smaller than 2 ft may be used if all edges are properly blocked.

[140] UBC Sec. 2513(a) states that diaphragms are to be considered incapable of supporting rotation (torsion). Also, see Sec. 126.

For example, wall heights must be less than 12 ft, and deflection/drift cannot exceed 0.5% (0.005 times story height). Other requirements apply.

The design of plywood shear walls is covered in UBC Sec. 2513, and in particular, Sec. 2513(e). Appendix L of this book, "Allowable Shear for Wind or Seismic Forces for Plywood Shear Walls with Framing of Douglas Fir-Larch or Southern Pine" (corresponding to UBC Table 25-K-1) gives the nailing schedule and other details for the design and analysis of such walls.

144 DIAPHRAGM RATIOS

Maximum aspect ratios of diaphragms are limited to the values listed in Table 18 (corresponding to UBC Table 25-I).

Table 18
Maximum Diaphragm Ratios
[UBC Table 25-I]

MATERIAL	HORIZONTAL DIAPHRAGMS Maximum Span-Width Ratios	VERTICAL DIAPHRAGMS Maximum Height-Width Ratios
1. Diagonal sheathing, conventional	3:1	2:1
2. Diagonal sheathing, special	4:1	3½:1
3. Plywood and particleboard, nailed all edges	4:1	3½:1
4. Plywood and particleboard, blocking omitted at intermediate joints	4:1	2:1

Reproduced from the 1991 edition of the *Uniform Building Code,* copyright ©️ 1991, with the permission of the publishers, the International Conference of Building Officials.

145 PLYWOOD DIAPHRAGM CONSTRUCTION DETAILS

A plywood sheet must be nailed continuously along its edges (*edge nailing*) and throughout its interior (*field nailing*) to achieve full development of the sheet. Table 17 (UBC Table 25-J-1) specifies the edge nail-to-nail spacing required to carry a particular shear load. The plywood is nailed to joists, blocking, and ledgers, as shown in Fig. 42.

All force elements must have a full transmission path across the diaphragm. Drag struts must frame into suitable walls or other collector elements. *Continuity ties* are required between adjacent edges of sheathing where edge nailing from the sheathing places the framing member below in cross-grain tension. (See Sec. 150.)

Figure 71 illustrates how ties can be used to transmit tension and compression forces through a perpendicular girder.

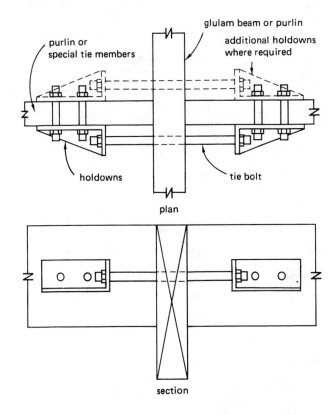

Figure 71 Typical Seismic Tie

146 BLOCKING

Blocking in accordance with UBC Sec. 2506(h) provides lateral support and prevents long joists from buckling—that is, "rolling out" or rotating out from under the members the joists support. When a joist buckles, its moment of inertia in the plane of support decreases significantly. Blocking is typically toe-nailed to the joists.

147 SUBDIAPHRAGMS

A *subdiaphragm*, as defined in UBC Sec. 2502, is "a portion of a larger diaphragm designed to anchor and transfer local forces to primary diaphragm struts and the main diaphragm." For example, in Fig. 72, the lateral forces from the masonry wall are transferred to the subdiaphragm through the anchor ties. The subdiaphragm span is the distance between the end shear wall and the center diaphragm strut. The anchor ties run the full span of the subdiaphragm. The subdiaphragm should be designed as if it acts alone in transferring the tie forces developed in the anchor to the shear wall and strut.

The maximum diaphragm ratios (see Sec. 144) apply to subdiaphragms. Therefore, the subdiaphragm depth limits the length of anchor ties required. The lengths of diaphragm struts and cross ties at diaphragm discontinuities are similarly limited.

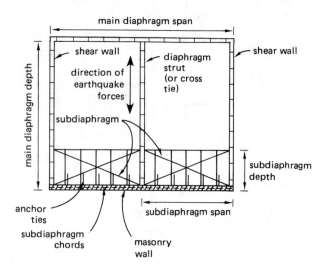

Figure 72　Subdiaphragm

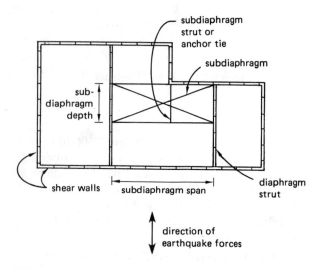

Figure 73　Subdiaphragm Limiting Strut Length

148 CONNECTOR STRENGTHS

Connectors, or fasteners, for wood are typically nails, lag bolts, and machine bolts. Connectors can fail in one of several ways. Connectors, particularly nails, can pull out; wood in shear connections can fail in bearing; connectors can fail in shear or bending. (Being much stronger than the wood pieces they connect, the connectors seldom fail in tension.)

It is generally unnecessary to deal with properties of the connectors such as yield strength in shear, longitudinal friction factor, and so on. The UBC provides tables of allowable loads for various types of connectors.[141] (See Table 19.)

Table 19

Allowable Lateral Loads for Nails
(Perpendicular to the Grain)
[UBC Table 25-G]

SIZE OF NAIL	STANDARD LENGTH (inches)	WIRE GAUGE	PENETRA-TION REQUIRED (inches)	LOADS (Pounds)[1 2 3]	
				Douglas Fir Larch or Southern Pine	Other Species
BOX NAILS					
6d	2	12½	1⅛	51	
8d	2½	11½	1¼	63	
10d	3	10½	1½	76	
12d	3¼	10½	1½	76	See U.B.C. Standard No. 25-17
16d	3½	10	1½	82	
20d	4	9	1⅝	94	
30d	4½	9	1⅝	94	
40d	5	8	1¾	108	
COMMON NAILS					
6d	2	11½	1¼	63	
8d	2½	10¼	1½	78	
10d	3	9	1⅝	94	
12d	3¼	9	1⅝	94	See U.B.C. Standard No. 25-17
16d	3½	8	1¾	108	
20d	4	6	2⅛	139	
30d	4½	5	2¼	155	
40d	5	4	2½	176	
50d	5½	3	2¾	199	
60d	6	2	2⅞	223	

[1]The safe lateral strength values may be increased 25 percent where metal side plates are used.

[2]For wood diaphragm calculations these values may be increased 30 percent. (See U.B.C. Standard No. 25-17.)

[3]Tabulated values are on a normal load-duration basis and apply to joints made of seasoned lumber used in dry locations. See U.B.C. Standard No. 25-17 for other service conditions.

---

[141]It is also important to recognize that connector forces used in wood-to-concrete and wood-to-masonry are limited by the strength of the concrete and masonry. Inasmuch as the wood is the weaker material, it seems logical that the wood provisions would determine the design, but there is no guarantee of this. Limitations for concrete and masonry are covered in UBC Secs. 2624 and 2406(h), respectively.

A. Nails   [UBC Sec. 2510(c)]

Table 19 (corresponding to UBC Table 25-G) gives the allowable lateral loads when a nail (box or common) is driven the specified distance perpendicular to the grain. Note that values may be increased 30% for use in wood diaphragm calculations, although it is conservative not to take the increase. The allowable loads are two-thirds of the table's values when nails are driven parallel to the grain. *Toe-nails* may be used to support up to five-sixths of the load. However, toe-nailing may not be used to connect diaphragms to ledgers in seismic zones 2, 3, and 4 [Sec. 2337(b)9D].

Table 20 (corresponding to UBC Table 25-H) gives the maximum withdrawal loads for nails driven perpendicular to the grain. Nails driven parallel to the grain of the wood are not permitted to support withdrawal loads. Other requirements for spacing and edge distances are given.

B. Bolts   [UBC Sec. 2510(b)]

Safe loads for bolts loaded in shear in wood-to-wood connections are given in Table 21 (corresponding to UBC Table 25-F). The shear force permitted in members used to connect wood to concrete or masonry is one-half of the double-shear values in the table for a wood member that is considered twice as thick as its actual thickness. The values may be increased by one-third for seismic loadings.

C. Ties

Allowable loads on proprietary structural ties must be given by the manufacturer of those ties. (See App. K.)

Table 20
Allowable Withdrawal Loads for Nails
[UBC Table 25-H]

KIND OF WOOD	SIZE OF NAIL									
	6d	8d	10d	12d	16d	20d	30d	40d	50d	60d
1. Douglas Fir, Larch	29	34	38	38	42	49	53	58	63	67
2. Southern Pine	35	41	46	46	50	59	64	70	76	81
3. Other Species	See U.B.C. Standard No. 25-17									

149 FLEXIBLE DIAPHRAGM TO WALL CONNECTION DETAILS

There are many acceptable ways that flexible diaphragms can be connected to shear walls. (There are even more methods of making connections to wood-framed walls.) All acceptable methods provide an unbroken path for the force to follow from the diaphragm to the soil. (Any detail where the joists, ledgers, or diaphragm merely sit on supports without connection is unsatisfactory. Toe-nailing and nailing subject to withdrawal are not permitted in seismic zones 2, 3, and 4 [UBC Sec. 2337(b)9D]).

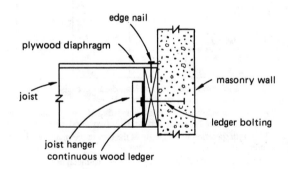

Figure 75   Joist-Wall Framing

Figure 75 shows a typical detail for a parallel wall into which the joist (which may also be called a purlin or other framing member term) ends the frame. The lateral force from the diaphragm travels from the plywood through the edge nails into the *ledger*. (A ledger may also be called a *nailer* or *sill*.) The force

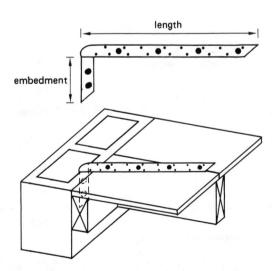

Figure 74   Typical Proprietary Ties

Table 21
Allowable Shear Loads in Bolts[1,2,4]
(Douglas Fir-Larch, California Redwood (close grain), and Southern Pine)
[UBC Table 25-F]

p = Safe loads parallel to grain in pounds
q = Safe loads perpendicular to grain in pounds

Length of Bolt in Main Wood Member[3] (in inches)		DIAMETER OF BOLT (IN INCHES)								
		⅜	½	⅝	¾	⅞	1	1⅛	1¼	1½
1½	Single p	325	470	590	710	830	945			
	Shear q	185	215	245	270	300	325			
	Double p	650	940	1180	1420	1660	1890			
	Shear q	370	430	490	540	600	650			
2½	Single p		630	910	1155	1370	1575			
	Shear q		360	405	450	495	540			
	Double p	710	1260	1820	2310	2740	3150			
	Shear q	620	720	810	900	990	1080			
3½	Single p			990	1400	1790	2135	2455	2740	3305
	Shear q			565	630	695	760	825	895	1020
	Double p	710	1270	1980	2800	3580	4270	4910	5480	6610
	Shear q	640	980	1130	1260	1390	1520	1650	1780	2040
5½	Single p				1950	2535	3190	3820	4975	
	Shear q				1090	1190	1300	1395	1605	
	Double p		1270	1990	2860	3900	5070	6380	7640	9950
	Shear q		930	1410	1880	2180	2380	2600	2790	3210
7½	Single p								3975	5680
	Shear q								1900	2185
	Double p			1990	2860	3890	5080	6440	7950	11,360
	Shear q			1260	1820	2430	3030	3500	3800	4370
9½	Single p									5730
	Shear q									2765
	Double p				2860	3900	5080	6440	7950	11,460
	Shear q				1640	2270	2960	3710	4450	5530
11½	Single p									
	Shear q									
	Double p				3900	5080	6440	7950	11,450	
	Shear q				2050	2770	3540	4360	6150	
13½	Single p									
	Shear q									
	Double p					5100	6440	7960	11,450	
	Shear q					2530	3310	4160	6040	

[1]Tabulated values are on a normal load-duration basis and apply to joints made of seasoned lumber used in dry locations. See U.B.C. Standard No. 25-17 for other service conditions.

[2]Double shear values are for joints consisting of three wood members in which the side members are one half the thickness of the main member. Single shear values are for joints consisting of two wood members having a minimum thickness not less than that specified.

[3]The length specified is the length of the bolt in the main member of double shear joints or the length of the bolt in the thinner member of single shear joints.

[4]See U.B.C. Standard No. 25-17 for wood-to-metal bolted joints.

Reproduced from the 1991 edition of the *Uniform Building Code,* copyright © 1991, with the permission of the publishers, the International Conference of Building Officials.

continues through the ledger bolts into the masonry wall and through the parallel wall to the foundation and ground. Such a connection will transfer forces parallel to the wall.

The framing method shown in Fig. 75 is not adequate (nor is it permitted [UBC Sec. 2337(b)9D]) for forces perpendicular to the wall because of cross-grain bending. (See Sec. 150.) Additional tension connection straps, as shown in Figs. 76 and 77, must be added [UBC Sec. 2337(b)9C]. The tension connection should be continuously nailed back (i.e., strapped) a considerable distance into the diaphragm to eliminate the high local tensile stress that would otherwise occur.

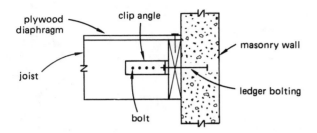

Figure 76 Joist-Wall Framing for Tension Forces

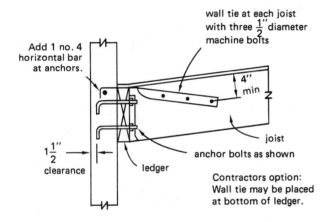

Figure 77 Ledger and Joist Tie

Figure 77 illustrates the typical details of a connection using both anchor bolts (for the ledger) and an embedded tie (for the joist[142]).

The framing for a wall parallel to the joists is shown in Fig. 78. The design (i.e., using a ledger) is basically the same as for the connections at the ends of the joists. (Notice that the detail as shown places the ledger in cross-grain bending. (See Fig. 79.) Straps would also be required between the diaphragm and wall.)

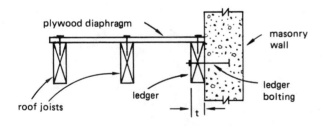

Figure 78 Joist-Parallel Wall Framing

The UBC [Sec. 2310] requires that (1) the connection between concrete and masonry walls and floor and roof

diaphragms must be designed to withstand the greater of seismic-induced forces or 200 pounds per lineal foot of wall, (2) the anchor spacing cannot exceed 4 ft unless the wall is designed to resist bending between anchors, and (3) the anchors must be grouted in place when they are embedded in hollow masonry blocks or cavity walls.[143] (See Sec. 110 for special provisions regarding the connector design force in the center of the diaphragm.)

150 CROSS-GRAIN LOADING

Cross-grain bending and *cross-grain tension* in ledgers and joists, as illustrated in Fig. 79, are not permitted [UBC Sec. 2337(b)9D].

151 FRAMING FOR DIAPHRAGM OPENINGS

It is not uncommon for a diaphragm to have openings for skylights, furnace flues, stairwells, and so on. The UBC requires in Sec. 2513(e)1B that such openings be completely blocked around the opening edges, with such blocking framing into joists and other structural members. The purpose of the framing is to redistribute shears from areas adjacent to the openings around, or past, the openings to other collection elements.

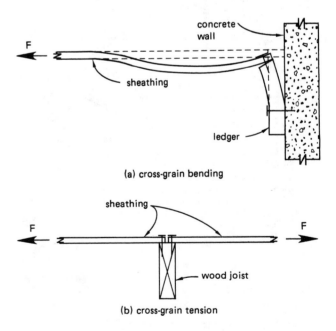

Figure 79 Cross-Grain Bending and Tension

---

[142] As Sec. 121 implies, the term *joist* may be replaced by *purlin* or other framing member term.

[143] It is usually easier to specify a 4-ft anchor spacing than to design the wall connection for bending.

An opening in the diaphragm imposes the following two requirements if the diaphragm is to operate correctly.

◇ The blocking around the perimeter of the opening must transfer the unequal loads on each side of the opening.

◇ The members around the perimeter of the opening must run from wall to wall, with tension straps used at corner connections to maintain continuity of the members. (See Fig. 80.) It is common to use double members for perimeter blocking. This eliminates the prohibited practice of using the diaphragm plywood to splice perimeter wood.

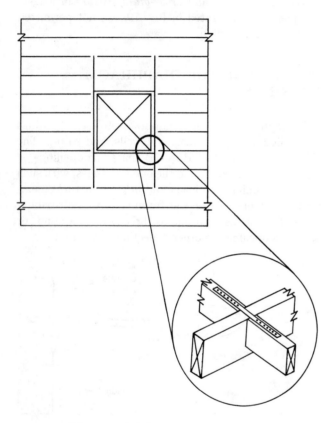

Figure 80   Framing an Opening

13

TILT-UP CONSTRUCTION

152 TILT-UP CONSTRUCTION DETAILS

A *tilt-up building* typically uses precast structural panels, a plywood diaphragm roof supported on wood joists or purlins, and either steel or glued-laminated (glulam) wood girders.

Analysis of tilt-up concrete shear walls is essentially the same as for cast-in-place concrete walls except that the panel-to-panel, panel-to-ceiling, and panel-to-floor details become critical. Shear between two panels must be developed by shear keys, dowels, or welded inserts. Contact joints are assumed to develop no strength in shear or in tension.

Tilt-up wall construction used in one-story industrial and commercial buildings fared poorly in the 1971 San Fernando and 1987 Whittier earthquakes. The main weaknesses were found in the connections, or *anchorages*, between the roof and walls, particularly between main girders, purlins, and joists and the walls, the connection of the perimeter wood *ledger* to the wall, and the nailing of the plywood diaphragm to the ledger. Basically, the walls moved outward, the girders and purlins detached from the ledgers, and the roofs fell to the ground.

Another problem with tilt-up construction occurs because each panel in a line (i.e., as part of a wall) is separate from the other panels and, therefore, resists a seismic load parallel to the panel in proportion to its relative rigidity. Since all of the panels are connected at their tops and bottoms, all will deflect the same amount. However, solid panels will resist the seismic load in shear (i.e., as a shear wall) while panels with large openings such as windows or doors will resist the seismic load in bending (i.e., as a beam).

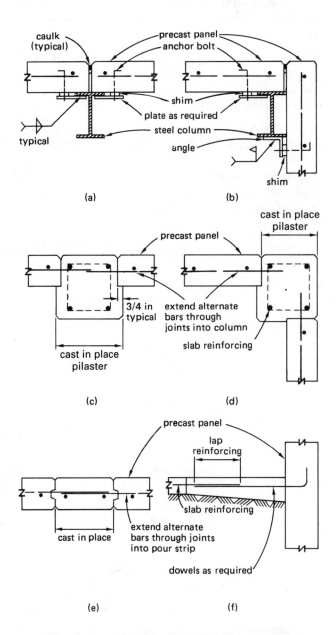

Figure 81   Details of Tilt-Up Construction Connections

Connections, such as those at weld plates, to shear walls should be numerous and regular, with a maximum spacing of approximately 4 ft, along the top of the shear wall, and such connections should be capable of transferring three times the expected lateral load. It is also necessary for each panel to be attached to the floor to counteract the overturning moment. (See Sec. 133.) These details will prevent the poor performance that has been experienced in some previous earthquakes.

Openings in tilt-up walls have become so numerous that a wall, with its openings and spandrels, is sometimes more like a frame. The 1990 Blue Book (Sec. 3A.3(f)) defines a *wall pier* as "... a wall segment with a horizontal length-to-thickness ratio between 2.4 and 6, and whose clear height is at least two times its horizontal length."

The definition does not give guidance as to when a wall is really a frame, nor was the definition picked up in the 1991 UBC. This will, undoubtedly, be the subject of future UBC provisions.

In the 1989 Loma Prieta earthquake, a number of "well-designed" tilt-up buildings experienced differential movement between the panel pilasters and the glulam roof beams that sat atop the pilasters. Greater attention should be given to detailing the embedment of the anchor bolts at the top of the pilaster to keep them from spalling the pilaster concrete.

14

SPECIAL DESIGN FEATURES

153 ENERGY DISSIPATION SYSTEMS

Various active and passive devices that reduce the magnitude or duration (or both) of the seismic force are in use or evaluation. These devices include active mass systems, passive visco-elastic dampers, tendon devices, and base isolation, and may be incorporated into the design when approved by the building official [UBC Sec. 2333(j)].

154 BASE ISOLATION

The base shear experienced by a structure is the product of the structure mass and the acceleration (i.e., $F = ma$). Little can be done to reduce the mass of a structure in an earthquake, but the acceleration can be reduced if the structure is not attached rigidly to its foundation. Application of this concept is known as *base isolation* or *decoupling*, and the connections between the structure and the foundation are known as *isolation bearings*. Base isolation is applicable to bridges as well as to buildings.[144]

In effect, the ground is allowed to move back and forth under the building during an earthquake, leaving the building "stationary." Since the building theoretically does not accelerate, it does not experience a seismic force. In most cases of base isolation, the building is partially constrained, but the concept is the same. This can be done by "skewering" the base isolator with a vertical rod surrounded by a clearance hole. It may be necessary to excavate a trench (or "moat") around the building to allow for differential movement.[145] Other nonstructural considerations include attaching utility service with flexible pipes and cables, as well as suspending elevator pits from the basement.

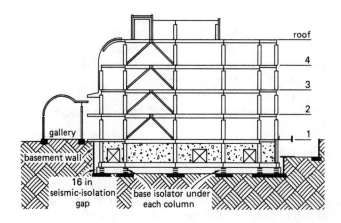

Figure 82   Base Isolation (Foothill Center, San Bernardino County, California)

---

[144]It is generally accepted that only bridges supported on box-girders, less than 300 ft long, and whose superstructures are supported at every pier (rather than being monolithic) are true candidates for base isolation. Longer monolithic bridges are already so flexible that their natural period is long enough to reduce stresses significantly. It is also usually cheaper to design the bridge pier foundations as moment-resisting members than to specify base isolation in new construction (as opposed to retrofitting old bridges). The first new bridge to use base isolation was the Sexton Creek Bridge near Cairo, Illinois, installed by the Illinois Department of Transportation. In 1986, the Metropolitan Water District (MWD) of Southern California used base isolation on its Santa Ana River crossing of the Upper Feeder pipeline. As of 1989, California's Department of Transportation (CALTRANS) had retrofitted five bridge structures.

[145]Every building is different, but a seismic isolation gap of approximately 12 to 16 inches should be used, although differential motion may need to be limited to less than the full gap size. After that, the deflection control element takes over and the building follows the earthquake motion. Refer to UBC Secs. 2337(b)11 and 2377(c)2.

Bearings consisting of elastomeric (e.g., neoprene) alone may be suitable for absorbing horizontal (thrusting) loads, but metal is incorporated into designs that support vertical loads. There are three primary methods of base isolation. These are supporting the building on (1) large ball bearings sandwiched between plates, (2) elastomeric bearings consisting of alternating layers of steel (or lead) and rubber, and (3) traditional structural expansion joints consisting of a layer of Teflon and a layer of rubber sandwiched between two steel plates. Combinations of these three methods (e.g., some bearings and some expansion joints) are desirable from a cost standpoint, since true bearings are costly. Initial testing has shown that such combinations work nearly as well as "pure" base isolation systems.

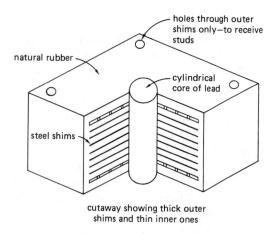

Figure 83   One Type of Isolation Bearing

Most base isolators consist of alternate layers of some elastomer (i.e., rubber) and steel plate. Another type of isolator is a Teflon-coated slider. Teflon sliders, though capable of isolation, do not provide significant damping. Therefore, they will most likely be used in hybrid systems in conjunction with high-damping rubber bearings.

The number of isolators necessary in an existing building depends on the type of foundation. A structure with a continuous foundation around its periphery may require hundreds of small isolators to spread out the building weight. However, a structure with a column-like support system might require only one isolator per column.

While base isolation is not yet in widespread use, there is evidence that the technique is successful.[146] Natural

building periods have been more than doubled, moving the structures into areas on the response spectrum of lower acceleration. (See Sec. 65.) In several recent earthquakes, most buildings experienced maximum accelerations well in excess of the ground acceleration. However, buildings using base isolation experienced accelerations from 25% to 50% lower than the ground acceleration. Building and bridge periods are 2 to 10 times larger than their original values.

It has been suggested that a building fitted with base isolation will experience a Richter magnitude 8 earthquake as a magnitude 5 or 5.5 earthquake.[147] Instrumentation in the four-story Foothill Center illustrated in Fig. 82 was operational during the April 1990 magnitude 5.5 earthquake epicentered approximately 7 mi away. (See App. C.) The base isolation system reduced the seismic forces by almost half. The foundation acceleration was 0.15 g but was only 0.08 g above the isolators. Acceleration was 0.16 g at the roof. The acceleration at two similar buildings approximately 6 mi from the epicenter was 0.13 g at the ground and 0.39 g at the roof.

Not everyone, however, is convinced that base isolation will stand the test of time. One concern is that no base-isolated building has experienced a truly significant earthquake. Some engineers are concerned that the elastomer will deteriorate, perhaps due to atmospheric ozone or other contaminants in the building, or that the bearing will "freeze up" after many years.

Another possible problem with base isolation derives from its very benefit—that of lengthening the natural period of the building. A natural soil period on the order of 3 or more seconds (as in cases where the soil is "soft") may coincide with the lengthened period of the building. This will produce resonance effects, such as those that occurred in the 1985 Mexico City and 1989 Loma Prieta earthquakes.

Base isolation is not suitable, economically or practically, for every building. One reason is that the base isolation bearings must all, within reason, be located at the same elevation. Buildings with footings that step down hillsides, for example, are poor candidates for base isolation.

---

[146]As of 1989, approximately only 12 buildings in the United States used base isolation. As of 1986, New Zealand had over 40 base-isolated buildings, and Japan had approximately 50 buildings and 80 bridges using base isolation. The first new building

to use base isolation was the 1983 Foothill Communities Law and Justice Center in San Bernardino, California. (See Fig. 82.) Base isolation has also been used in the "seismic-proofing" of existing structures, such as the Salt Lake City and County Building (which opened in 1989), since it is one of the few methods that does not require extensive exterior work.

[147]Time will tell.

Only a short time ago, seismic isolators were "outside" the provisions of the UBC and were considered exotic and experimental. Now they are installed routinely as retrofit devices to add seismic protection to buildings and bridges.

Base isolation is now specifically permitted (subject to the approval of the building official) [UBC Sec. 2333(j)].

A new appendix in the 1991 UBC covers the design of seismic-isolated buildings and nonbuilding structures in detail. Appendix 23, Div. III, "Earthquake Regulations for Seismic-Isolated Structures," essentially replaces the provisions of Chapter 23. This new appendix defines the design lateral force for buildings and parts/portions based on a design displacement.

Other provisions are parallel to Chapter 23. Specifically given are conditions for using static and dynamic analyses, methods for distributing the lateral force to the stories, drift limitations, and overturning provisions.

Design of base isolation systems is covered in the UBC [App. 23, Div. III, Secs. 2370–2381, "Earthquake Regulations for Seismic-isolated Structures"].

155 DAMPING SYSTEMS

Passive and active damping systems, like base isolation, are in their infancy, at least in terms of large-scale use. These systems increase the damping ratio of the building and, in so doing, decrease the amplitude of swaying. Some involve moving blocks and counterweights, while others, such as *passive visco-elastic dampers* or *friction dampers*, are not much more than large shock absorbers.[148] Those that require power for motors and information from sensors for computers are known as *active systems*; those that do not are *passive systems*.

Already used in some high-rise buildings to reduce wind drift, *active mass dampers* are nothing more than multi-ton blocks, usually of concrete or steel, suspended like a pendulum by a cable or mounted on tracks in one of the building's upper stories. When the wind or an earthquake makes the building sway, a computer sensing the motion signals a motor to move the weight in the opposite direction, thereby minimizing or neutralizing the motion.

Only a specific size of block will work in a building, because the weight of the block depends on the building's weight, location of the block, lag time, and mode

---

[148]Friction dampers have more in common with devices used to absorb coupling shocks in railway rolling stock than with automobile shock absorbers.

to be counteracted. Therefore, the mass is "tuned" to the structure, and the systems are also known as *tuned mass dampers (TMD)*.

Active tendon and *active pulse systems* are similar, except that the building is moved by hydraulic pistons in the foundation or between stories instead of by a mass at the top. The energy pulse usually only needs to be applied once or twice each building motion cycle.

These devices typically reduce the lateral forces by one-third to one-half while increasing the building weight approximately 1%. They are also suitable for torsion control when placed off-center in a structure.

The most significant drawback to active systems is the fact that they require external power not only for the computer but also for the motors driving the masses. Further, the large masses currently in use ride on oil bearings that take up to four minutes to pressurize. Thus, while active systems are useful in reducing drift during a predicted and slowly-increasing windstorm, such devices are not yet substitutes for proper seismic design.

Active damping systems remain largely experimental. An exception is located in Tokyo in the Kyobashi Seiwa Building known for its extraordinary shape (11 stories high and only 13 feet wide).

Most damping devices installed in buildings are passive. (The best-known passive system is located on the top floor of the 59-story Citicorp Building in Manhattan, where a 400-ton, concrete tuned block is located.)

The term "ADAS" refers to "Added Damping and Stiffness" elements. An ADAS element is a passive damping system that generally consists of a combination of steel plates and spacers. The plates are bolted to structural bracing at the plate tops and bottoms. As the top and bottom structural bracing members displace relative to one another, the ADAS plates bend (i.e., yield) and dampen vibrations. The advantage of ADAS elements over conventional damping systems or shock absorbers is that ADAS elements contain no moving parts and require no maintenance.

156 ARCHITECTURAL CONSIDERATIONS

Due to planned yielding, the inter-story deflections in a major earthquake will be several times larger than the elastic deflections that are calculated from the base shear equation in the UBC. Therefore, damage to architectural (i.e., nonstructural) items is likely. Even for lesser-magnitude events, however, nonstructural items must be properly detailed.[149]

Proper architectural detailing means providing proper clearances for exterior cladding, glazing (i.e., glass), wall finishes (e.g., marble veneers), interior partitions, and wall panels. Chimneys in residential buildings must be properly reinforced internally and securely strapped to the building at the roof line. Some elements can be floated free—that is, they can move independently of the building. Proper attention must be given to the connection of these elements to the building.

To ensure that the occupants are able to get out of the building, doors should be designed to remain functional.

Floor coverings must be capable of three-dimensional movements.

All elements capable of falling and causing damage or injury must be rigidly attached to structural members. For example, suspended ceilings (e.g., tee-bar) and lights in drop-in ceilings must be tied to ceiling members above. Partitions, particularly those that do not run floor-to-ceiling, require special attention.

Columns in traditional moment-resisting frames are typically spaced 3–7 m (10–20 ft) apart. This spacing presents a challenge to architects when they try to provide unobstructed occupant space in high-rise buildings. Designers who want exterior column spacings greater than this must use other techniques to increase the strength of their buildings. Use of high-strength concrete, braced cores, ductile frames, ductile outrigger framing, and bandages are some of the techniques used.

Most trusses contain numerous axial members that are arranged in triangular sections. These triangular sections, though effective, make it difficult to include corridors (when in the core of the building) and windows (when on the perimeter of the building) in the design.

With the ability to create efficient moment-resisting joints, an increasing number of engineers are designing trusses comprised of rectangular sections. A bridge-like truss comprised of rectangular sections with moment-resisting joints is known as *quadrangular girders, open-web girders* or *trusses, ductile frames,* or *Vierendeel girders*. In addition to providing unobstructed access through the openings in the truss, these structures are economical to build.

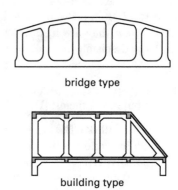

bridge type

building type

Figure 84   Quadrangular Girders

In-fill panels can either be given sufficient clearance so that they are not crushed by the deflection of adjacent structural elements, as in a panel forming a wall between two columns, or short sections of the panel (usually at its vertical edges) can be designed to be flexible or much weaker and replaceable.[150] Buttresses—short panel tees or wall returns—should be used to prevent the panel from falling over.

Computer room floors that are raised on pedestals and placed on unbraced stringers have fared poorly in past earthquakes.

Equipment such as air-conditioning devices, motors, pumps, tanks, piping, and air ducts must be securely bolted to their foundations to prevent sliding and overturning. Mere use of clamps or clips to prevent overturning is inadequate because the equipment can slide out from under the clips during an earthquake.

157 UPGRADING EXISTING CONSTRUCTION

Structures built in prior years can be upgraded with new features to make them more resistant to seismic forces. Such an activity is known as *retrofitting*. Bridges (some of which were heavily damaged in the 1971 San Fernando earthquake) can be fitted with cable restraints and increased capacity shear keys to restrict longitudinal motion.

Nonductile concrete columns, built before the ductility requirements were added to the UBC in 1973, with rectangular cross sections have been upgraded by wrapping

---

[149]This sentence probably should begin "Particularly for lesser-magnitude events" because who wants architectural damage in the more common small events?

[150]Care must be taken when using flexible materials that they remain flexible indefinitely. Foamed polyethylene and polysulfide products appear *not* to meet this requirement.

PROFESSIONAL PUBLICATIONS, INC. ● Belmont, CA

them in steel plates. It has been proposed that columns with round or nonrectangular cross sections be wrapped with steel or composite fiber wire to give them greater ductility.

Carbon fibers are probably too brittle to be used to wrap columns as a means of achieving spiral strand confinement. However, CALTRANS is investigating the use of Kevlar fibers (the same material used in most bullet-proof vests) to wrap columns.

Steel jackets are comprised of two semicircular portions welded up the seams. Grout is injected into the space between the jacket and column. Thickness of the steel jacket depends on the loads expected.

The proper application of flat steel plates ("Vierendeel bandages," "Vlasovian bandages," or just "bandages") at selected exterior locations on a structure can increase resistance to seismic forces. Bandages do this by increasing the torsional moment of inertia of the structure (when the bandages are connected to core framing members) and by providing alternate vertical load paths (when the bandages are connected to load-carrying members at different levels). Inasmuch as only portions of a structure are covered, bandages are not effective in containing concrete. Therefore, this technique should not be confused with the encasement of concrete columns in steel jackets.

Belting, as a means of upgrading an unreinforced masonry building, is a technique that extends long steel rods horizontally from corner to corner, where the rods are attached to plates on the building's corners. The resulting "girdled" nature of the building is supposed to keep the building's walls from pulling away from the building core during an earthquake. For this technique to work, the rods must placed relatively close together, and they must extend all the way up the building.

There are no codes specifically governing the methods of retrofitting or upgrading existing structures, and no one really knows how successfully retrofitted structures will fare in an earthquake. Approximately 40% of the unreinforced masonry buildings that had been retrofitted sustained damage in the 1987 Whittier earthquake. During the 1989 Loma Prieta earthquake, a four-story unreinforced masonry building in San Francisco that had been seismically retrofitted by belting suffered major seismic damage.

15
PRACTICE PROBLEMS

EARTHQUAKES IN GENERAL

1. What is the lowest acknowledged numerical Richter magnitude that would identify a major earthquake?

Answer

This is an ambiguous question since different people would interpret the word *major* differently. In general, a moderate earthquake would have a Richter magnitude of 5, a strong earthquake would have a magnitude of 7, and a great earthquake would have a magnitude of 8 or higher. (See Sec. 18.)

2. What is the theoretical upper numerical limit on the Richter magnitude scale?

Answer

There is no theoretical upper limit on the Richter magnitude scale. (See Sec. 18.)

3. What does the Richter magnitude scale measure?

Answer

It is not clear what is intended by the word *measure*. The Richter apparatus *detects* earth movement. The numerical magnitude *describes* earthquake strength. The numerical value *represents* a measure of energy release on a logarithmic scale. (See Sec. 18.)

BUILDINGS

1. What is the difference between *stiffness* and *rigidity* as used in seismic consideration?

Answer

Stiffness is the force that is applied to deflect a structure a unit amount in a given direction. *Rigidity*—strictly *relative rigidity*—is a normalized stiffness. Whereas the stiffness of a single member can be used in numerical calculations, rigidities can only be used when forces are being distributed among several members. (See Secs. 47 and 48.)

2. What is the difference between *ductility* and *flexibility* as used in seismic consideration?

Answer

Flexibility is the reciprocal of *stiffness*. It is the elastic deflection obtained when a unit force is applied. (See Sec. 47.) *Ductility* is the ability of a material to distort and yield without fracture or collapse. (See Sec. 69.) Since flexibility deals with elastic deformation and ductility deals with inelastic deformation, there is little connection between the two concepts.

3. What is the relationship between *rigidity* and the variables of pier height, depth, and thickness?

Answer

Roughly, *rigidity* is proportional to the first power of thickness and to the cube of pier depth and is inversely proportional to the cube of pier height. (See Sec. 48.)

4. What is *ductility*?

Answer

Ductility is the ability of a material to distort and yield without fracture or collapse. (See Sec. 69.)

5. What is the *ductility factor*?

Answer

The *ductility factor* of a material is the ratio of its strain energy at fracture to its strain energy at yield. There are other similar and related definitions. (See Sec. 70.)

6. What factors influence the ductility factor?

Answer

From a metallurgical perspective, temperature and previous stress-strain history influence the ductility of a ductile material such as steel. The higher the temperature, the greater the ductility. The more the material has been worked or stressed in previous cycles or events, the more brittle (the opposite of ductile) it becomes. From a structural perspective, ductility depends on the type of construction (i.e., steel or concrete), the structural system, the quality of construction, the detailing, and the redundancy. (See Secs. 69 and 70.)

7. What is the minimum recommended ductility factor?

Answer

It is not possible to specify a minimum recommended ductility factor exactly because it depends on the type of structure, construction material used, intended use of the structure, and many other factors. However, the ductility factor should be well in excess of 1.0 and seems to be no less than 2.5 for modern structures. (See Secs. 69 and 70.)

8. What is *ductile framing*?

Answer

In its simplest interpretation, a structure with ductile framing will not collapse even though its structural frame has sustained significant distortion, misalignment, and other yielding damage. (See Sec. 139.)

9. What is the principle reason for specifying a minimum ductility factor?

Answer

The principle reason for specifying a minimum ductility factor is to obtain a *ductility margin* (i.e., the ductility between yield and collapse) sufficient to ensure survivability in a design earthquake.

10. Why will a theoretical analysis of elastic response of a structure usually overestimate the stresses resulting from an earthquake?

Answer

A structure will not behave totally elastically during an earthquake. Local yielding at high stress locations reduces the seismic energy, i.e., the energy of oscillation, initially present in the structure.

11. Describe the two components of *drift*.

Answer

Shear drift is the sideways deflection of a building due to lateral (sideways) loads. *Chord drift* is the sideways deflection due to axial (vertical) loads. (See Sec. 74.)

12. What is the P-Δ effect?

Answer

The P-Δ effect is an additional column bending stress caused by eccentric vertical loads. (See Sec. 75.)

13. How are drift and the P-Δ effect related?

Answer

When a structure drifts, its vertical loads become eccentric. The eccentric loading increases the column stress, and the stress increase is called the P-Δ effect. (See Secs. 74 and 75.)

14. What is the *natural period* of a building?

Answer

The *natural period* of a building is the time it takes the building to complete one full swing in its primary mode of oscillation. (See Sec. 41.)

15. What does the term *redundancy* mean as it is used in the context of modern high-rise buildings?

Answer

Redundancy, as used in a seismic context, is synonymous with *distributed excess capacity* and *multiple stress paths*. A *redundant design* has a safety factor, but the converse statement is not necessarily true. For example, if a vertical 100-kip load is supported by a single column having a 120-kip capacity, the design will have excess capacity ·but no redundancy since the structure will collapse if the column fails. If the 100-kip load is supported by 12 columns, each with a 10-kip capacity, the design will have both redundancy and excess capacity.

16. Has the recent trend in high-rise buildings been toward increased or decreased redundancy? Why?

Answer

Redundancy is increasingly seen as a crucial characteristic of high-rise designs. Multiple redundant stress paths greatly increase the reliability of a structure.

Since building members (e.g., columns, girders, and shear walls) and details (i.e., column-girder joints) do not always behave as intended (due to our meager knowledge of the behavior of so-called ductile designs, design or construction errors, and higher-than-expected loading), the design should allow for the unintended loss of all capacity in a small fraction of the members. The tolerable loss of this excess capacity is the principle of redundant design.

17. What causes *torsional shear stress*?

Answer

Torsional shear stress occurs when an earthquake acts on a structure whose centers of mass and rigidity do not coincide. (See Sec. 76.)

18. What is *negative torsional shear stress*?

Answer

Negative torsional shear stress is the torsional shear stress on one side of a structure that is opposite in sign to the shear stress induced by the base shear. (See Sec. 77.)

19. How should negative torsional shear stress be treated?

Answer

Inasmuch as the direction of an earthquake is not known in advance, negative torsional shear stress should be disregarded—it should not be used to decrease the size of a wall or other member. (See Sec. 77.)

20. What is the *Rayleigh method* and where would it be used?

Answer

There are at least two vibration analysis procedures that are known as the Rayleigh method. The first is a rule of thumb that is used to correct for the assumption of massless springs in vibration problems. (A fraction—usually one-third—of the spring mass is added to the oscillating mass.) The second procedure is a method determining the mode shape of a multiple-degree-of-freedom system through an iterative process. (See Sec. 63.)

21. Explain *critical damping*.

Answer

Critical damping is the amount of structural damping that causes oscillation to die out and return to the equilibrium position faster than any other amount of damping. (See Sec. 51.)

22. What is the *damping ratio*?

Answer

The *damping ratio* is the ratio of the actual damping coefficient to the critical damping coefficient. (See Sec. 51.)

23. What is the practical range of damping ratios?

Answer

Damping ratios of typical buildings range from approximately 0.02 for steel-frame construction to around 0.15 for wood-frame construction. (See Sec. 53.)

24. To what extent does damping affect the natural period of vibration of a structural frame?

Answer

Damping increases the actual period of vibration slightly, compared to the natural period of vibration. However, even with highly-damped structures, the increase is usually 1% or less. Therefore, the natural period is used in the UBC calculations and the effect of damping is disregarded. (See Sec. 54.)

25. What is a *response spectrum*?

Answer

A *response spectrum* is a graph of the effective peak acceleration that a building experiences plotted as a function of the building's natural period. (See Sec. 65.)

26. How does the *portal method* deal with the effects of column lengthening and shortening?

Answer

The *portal method* disregards changes in column length. (See Sec. 135.)

STRUCTURAL SYSTEMS

1. Distinguish between a *moment-resisting frame* and a *special moment-resisting frame*.

Answer

A *moment-resisting frame* has rigid connections between members (e.g., between girders and columns) such that moments applied to columns are partially resisted by girder bending and vice versa. With sufficiently large moments, however, even the elastic capacity of such structural systems that share loads between girders and columns can be exceeded. The integrity and load-carrying ability of a *special moment-resisting frame* (previously known as a *ductile moment-resisting frame*) will remain intact even after yielding has been experienced. (See Sec. 96.)

2. What are *bearing wall systems* and *box systems*?

Answer

Box system is another name for *bearing wall system*. A bearing wall system relies on shear and load-bearing walls to carry dead, live, and seismic loads. (See Sec. 96.)

3. Generally speaking, which is more likely to have a smaller natural frequency of vibration, a steel moment-resisting frame or a concrete moment-resisting frame, given equal heights and moments of inertia?

Answer

While the performance of both steel and concrete frames are similar in this regard, some theoretical generalizations are possible. The steel frame may be slightly more flexible (smaller stiffness), producing a smaller frequency and longer period. Also, the concrete frame may have greater mass, producing a smaller frequency and longer period. (The effect of damping on the period is minimal.) The UBC equation for period (see Eq. 64) clearly indicates that steel buildings generally are expected to have longer periods (smaller frequencies). (See Sec. 94.)

4. Which is more likely to have a larger damping ratio, a steel or concrete moment-resisting frame?

Answer

Concrete construction generally has a greater damping ratio. (See Sec. 53.)

5. Are plastic hinges designed in columns, in girders, or in both?

Answer

A *plastic hinge* forms when a member yields. The yielding of a girder or of a girder-column joint may produce distortion, floor and roof sagging, and misalignment without collapse. The yielding of a column, however, may lead to structural collapse. Therefore, unlike girders, columns should not be designed to form plastic hinges.

6. What is the structural system called that does not have a complete vertical load-carrying space frame?

Answer

Bearing wall systems, or box systems, use walls, not frame members, to carry the vertical loads. (See Sec. 96.)

UBC

1. Describe how base shear is calculated according to the UBC equation.

Answer

The base shear is calculated as an equivalent static loading based on a fraction of the structure weight. Terms are included to account for the seismic zone, building occupancy, building period, soil type, and structural system. (See Sec. 90.)

2. Which of the terms in the base shear equation can be equal to 1.0?

Answer

The importance factor (I) can have a value of 1.0. It is also possible for the C coefficient to have a calculated value of 1.0. The building period (T) and site coefficient (S) can have values of 1.0, but these are not in the base shear equation. (See Secs. 92, 93, and 94.)

3. In the application of the base shear formula, what factor of R_w should apply if a structure is a steel eccentric-braced frame in one direction and a concrete shear-wall building in the orthogonal direction?

Answer

This is a trick question. Since the structure's performance is analyzed independently for earthquakes in the two orthogonal directions, the structure is treated as an eccentric-braced frame building ($R_w = 10$) for an earthquake in one direction and as a shear-wall building ($R_w = 6$) for the orthogonal direction. (See Sec. 96.)

4. Explain in general terms how the base shear is distributed in a horizontal plane to the various resisting elements.

Answer

The base shear is distributed to the resisting elements in proportion to their rigidities. (See Sec. 48.)

5. Draw the seismic force diagram acting on a multistory building.

Answer

Refer to Fig. 34(a), Sec. 105, or Fig. 35, Sec. 108.

6. Draw the cumulative shear diagram acting on a multi-story building.

Answer

Refer to Fig. 34(b), Sec. 105.

7. What is the UBC building height limit for ordinary shear-wall construction in seismic zones 3 and 4?

Answer

Shear wall construction may not be used in buildings higher than 160 ft. (See Table 14, Sec. 96.)

8. Which has a smaller R_w value, a steel or concrete special moment-resisting frame?

Answer

Steel and concrete structures with special moment-resisting frames have the same R_w value, (12). (See Table 14, Sec. 96.)

9. What possible values can R_w take on for a moment-resisting frame?

Answer

R_w can have values of 12 (SMRF of steel or concrete), 8 (concrete IMRF), 6 (steel OMRF), and 5 (concrete OMRF). (See Table 14, Sec. 96.)

10. What R_w value would be used for (a) a large football grandstand with bleachers and (b) a tall vertical tank supported on a raised platform supported by a braced framework?

Answer

(a) A football grandstand with bleachers is a self-supporting nonbuilding structure falling under the jurisdiction of the local building official. Its mass can be considered to be lumped at the various spectator levels, and the supporting system continues between floors. Therefore, it is covered by the UBC. However, it is not specifically mentioned in Table 16. (See Sec. 111.) It would probably have an R_w value of 4.

(b) The framework-supported vertical tank is specifically mentioned in the Blue Book commentary as a case not covered at all by the code provisions. The combined tall structure and frame base does not meet the requirement of a distributed supporting system, and mass cannot be considered lumped at various levels. (See Sec. 111.) Therefore, no R_w can be assigned.

11. What is the absolute UBC limitation on drift?

Answer

The key word in this question is "absolute." There is no absolute limitation on drift in the UBC, as any drift that can be shown to be "tolerable" is permitted. Other limitations apply, however, when drift is not tolerable. (See Sec. 107.)

12. What percentage of the seismic load should be carried by a special moment-resisting frame in a building taller than 160 ft?

Answer

All of the seismic load is carried by a special moment-resisting frame. (See Sec. 96.) There is no height limit for special moment-resisting frames.

13. What percentage of the live load in a warehouse should be added to the dead load when calculating base shear?

Answer

A minimum of 25% of the warehouse live load should be added. (See Sec. 97.)

14. What are the various seismic zones in California?

Answer

California has two seismic zones: 3 and 4. (See Sec. 35.)

15. What are the approximate maximum accelerations in the California seismic zones?

Answer

Referring to Table 5 (Sec. 35), the approximate maximum accelerations in zones 3 and 4 are 0.33 g and 0.50 g, respectively. These roughly correspond to the zone factors, Z, in Table 11 (Sec. 91). Strictly speaking, the question cannot be answered without specifying the period of time (i.e., the recurrence interval) in which the earthquakes occur.

CONCRETE AND MASONRY STRUCTURES

1. What are the possible modes of failure due to seismic forces if the lateral force-resisting system of a high-rise building is constructed of reinforced concrete?

Answer

(See Sec. 139.) This question does not specify whether the concrete is specially reinforced or whether the concrete is used in a frame or shear-wall structure. In general, a reinforced concrete frame will have failed if the concrete spalls or crushes before plastic yielding of the steel reinforcing occurs, or if the steel reinforcing is stressed plastically. Failure can be expected to occur:

(a) at the ends of well-designed columns when there is insufficient shear resistance (i.e., such that the column breaks out of its supports)

(b) in poorly designed columns with insufficient confinement

(c) in shear walls due to inadequate vertical reinforcing

(d) at construction joints due to inadequate bonding between members

(e) in beams due to inadequate shear reinforcing

(f) in columns due to excessive drift and overturning moment

2. What are the most important considerations in achieving ductility in concrete frames?

Answer

The most important considerations are confinement and continuity. (The steel in confined, or specially reinforced, concrete should yield before the concrete crushes.) Adequate bonding between steel and concrete must be ensured. Steel must be capable of developing its full tensile strength. Members must be adequately tied together at joints. (See Sec. 139.)

3. What is meant by *confined concrete*?

Answer

Confined concrete is also called *ductile concrete* or *specially reinforced concrete*. The steel in ductile concrete will yield before the concrete crushes. This enables the concrete member to develop its full compressive strength without yielding. (See Sec. 139.)

4. Why is concrete confined at joints and in members?

Answer

The confining steel in ductile concrete enables the concrete to develop its full compressive strength without yielding. (See Sec. 139.)

5. What are some of the construction methods used to ensure ductile behavior of concrete?

Answer

To confine concrete, columns are spiral wrapped at closer intervals and additional hoops are used at joints and other locations. Continuity of reinforcement is achieved by special attention to splices. Special attention is given to reinforcement of shear walls. Hooks, ties, stirrups, and hoops are detailed to prevent pull-out. (See Sec. 139.)

6. With regard to resistance to seismic forces, which is better in steel-reinforced concrete columns, spiral ties or horizontal ties? Why?

Answer

Spiral transverse reinforcement is preferred, but it may not be possible to use it. From a construction standpoint, spiral reinforcement for smaller columns is easier to form in the field. From a seismic standpoint, spiral reinforcement provides slightly better confinement. Larger columns, however, cannot be wrapped in the field, and individual factory-fabricated ties must be used.

7. What is the function of the spiral and individual ties used in a concrete column?

Answer

Ties confine the concrete and keep it from crushing. (See Sec. 139.)

8. Do special reinforcement hoops replace regular ties in beams and columns?

Answer

Special reinforcement (primarily in the form of additional hoops) is used in addition to regular beam stirrups and column ties. In beams, special reinforcement is required at points of expected yielding (i.e., at plastic hinges). In columns, hoops are required at column-girder connections. (See Sec. 139.)

9. What is the effectiveness of stirrups in deep concrete beams?

Answer

After inclined cracks form at the ends of deep beams, the load is carried in a "tied arch" configuration that has considerable remaining strength. Stirrups in the center of a deep beam are not particularly effective.

10. What is the maximum permitted shear stress (in pounds per square inch) for a masonry wall with special inspection?

Answer

The maximum shear stress is 35 psi when in-plane flexural reinforcement is present. The maximum shear stress is 75 psi when steel takes all of the shear stress. (See Ftn. 114.)

STEEL STRUCTURES

1. What are the possible modes of failure due to seismic forces if the lateral force-resisting system of a high-rise building is constructed of steel?

Answer

In general, steel will be considered to have "failed" if it yields. (See Sec. 140.)

2. Where is a steel-framed building with a properly designed special moment-resisting frame most likely to yield in an earthquake?

Answer

Yielding and formation of plastic hinges in a steel structure can be expected at points where the moments are greatest, such as at girder ends and mid-spans and at column-girder connections. Columns can buckle due to bending and eccentric effects. Flanges and webs of members can buckle from local stresses and fail from fatigue loading. (Girder-column connections should not fail, however, through weld and bolt failure. All connections should be able to sustain the full plastic moment of connected members.)

SOILS AND FOUNDATIONS

1. Given a soil engineering report, what would you propose in order to improve the seismic response of the building?

Answer

Devastating resonance effects can be avoided if the natural building period does not coincide with the site period. (See Secs. 39, 41, and 57.) If the site has fine sand, soil liquefaction effects must be considered. (See Sec. 40.)

DESIGN PROBLEMS

1. The top plate of the structure shown is rigid. A 10-kip load is applied at the plate. What are the resisting shears in each column?

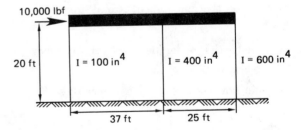

Solution

The total rigidity is proportional to the sum of the moments of inertia. The total is

$$I_{total} = I_1 + I_2 + I_3$$
$$= 100 + 400 + 600 = 1100$$

The load carried by the first (left) column is

$$V_1 = (10 \text{ kips}) \left(\frac{100}{1100} \right) = 0.91 \text{ kips}$$

The load carried by the second (middle) column is

$$V_2 = (10 \text{ kips}) \left(\frac{400}{1100} \right) = 3.64 \text{ kips}$$

The load carried by the third (right) column is

$$V_3 = (10 \text{ kips}) \left(\frac{600}{1100} \right) = 5.45 \text{ kips}$$

2. A power transformer is mounted on a pedestal. The structural adequacy for the seismic environment is to be demonstrated. The horizontal component of the seismic force is to be defined by the El Centro north-south elastic response spectra. If the lateral frequency of the system exceeds 30 Hz, the response is defined as being equivalent to a 0.5 g static load. The vertical response is defined as 50% of the horizontal response, and both occur simultaneously. Free lateral vibration tests on the structure show that the ratio of successive swings of the structure is 0.882. Assume that the pedestal mass and shear flexibility, foundation stiffness, and torsion can be neglected. Specific properties are shown in the figure.

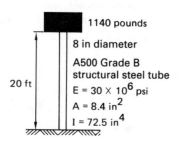

20 ft

1140 pounds

8 in diameter

A500 Grade B
structural steel tube
E = 30 × 10⁶ psi
A = 8.4 in²
I = 72.5 in⁴

(a) Calculate the undamped lateral frequency and period.

(b) Calculate the undamped vertical frequency and period.

(c) Explain how the assumptions affect these frequencies and periods.

(d) What is the logarithmic decrement for damping of lateral vibrations?

(e) What is the damping ratio?

(f) What is the lateral acceleration?

(g) What is the vertical acceleration?

(h) What is the base shear?

(i) Are the assumptions used to calculate the frequencies conservative with respect to the lateral load?

(j) Assume that the pedestal fails by bending. Does the structure have high ductility?

Solution

(a) From Table 7, the lateral stiffness of a vertical cantilever is

$$k = \frac{3EI}{h^3} = \frac{(3)\left(3 \times 10^7 \, \frac{\text{lbf}}{\text{in}^2}\right)(72.5 \text{ in}^4)}{(20 \text{ ft})^3 \left(12 \, \frac{\text{in}}{\text{ft}}\right)^2}$$

$$= 5664 \text{ lbf/ft}$$

The mass in slugs (equivalent to lbf-sec²/ft) can be calculated from the mass in pounds by dividing by the gravitational constant, g_c.

$$m_{\text{slugs}} = \frac{m_{\text{lbm}}}{g_c} = \frac{1140 \text{ lbm}}{32.2 \, \frac{\text{ft-lbm}}{\text{lbf-sec}^2}}$$

$$= 35.4 \text{ slugs}$$

From Eqs. 29 and 30, the natural period for lateral vibrations is

$$T_{\text{lateral}} = \frac{2\pi}{\omega} = 2\pi \sqrt{\frac{m}{k}}$$

$$= 2\pi \sqrt{\frac{35.4 \text{ slugs}}{5664 \, \frac{\text{lbf}}{\text{ft}}}}$$

$$= 0.497 \text{ sec (say 0.5 sec)}$$

From Eq. 28, the frequency is

$$f = \frac{1}{T} = \frac{1}{0.5 \text{ sec}} = 2 \text{ Hz}$$

(b) The deflection due to a compressive load is

$$\Delta L = \frac{FL}{AE}$$

The stiffness is the force per unit deflection.

$$k = \frac{F}{\Delta L} = \frac{AE}{L}$$

$$= \frac{(8.4 \text{ in}^2)\left(30 \times 10^6 \, \frac{\text{lbf}}{\text{in}^2}\right)}{20 \text{ ft}}$$

$$= 1.26 \times 10^7 \text{ lbf/ft}$$

Again from Eqs. 29 and 30, the natural period for vertical vibrations is

$$T_{\text{vertical}} = \frac{2\pi}{\omega} = 2\pi \sqrt{\frac{m}{k}}$$

$$= 2\pi \sqrt{\frac{35.4 \text{ slugs}}{1.26 \times 10^7 \, \frac{\text{lbf}}{\text{ft}}}}$$

$$= 1.05 \times 10^{-2} \text{ sec}$$

From Eq. 28, the frequency is

$$f = \frac{1}{T} = \frac{1}{1.05 \times 10^{-2} \text{ sec}} = 95 \text{ Hz}$$

(c) It is assumed that all flexibility comes from the tube. The flexibility of other elements in the structural system decreases the stiffness. This increases the period. It is also assumed that the tube is massless. Increasing the vibrating mass also increases the period.

(d) From Eq. 35, the logarithmic decrement, δ, is

$$\delta = \ln\left(\frac{x_n}{x_{n+1}}\right) = \ln\left(\frac{1}{0.882}\right) = 0.126$$

(e) From Eq. 35, the damping ratio is solved directly.

$$\delta = 0.126 = \frac{2\pi\xi}{\sqrt{1-\xi^2}}$$

$$\xi = 0.02 \ (2\%)$$

(f) From Fig. 25 with 2% damping (determined in part (e)),

$$S_d = 2.2 \text{ in}$$
$$S_v = 28 \text{ in/sec}$$
$$S_a = (0.9 \text{ g})\left(32.2 \frac{\text{ft}}{\text{sec}^2\text{-gravity}}\right)$$
$$= 29 \text{ ft/sec}^2$$

(g) The vertical response is 50% of the horizontal response.

$$S_d = (0.5)(2.2 \text{ in}) = 1.1 \text{ in}$$
$$S_v = (0.5)\left(28 \frac{\text{in}}{\text{sec}}\right) = 14 \text{ in/sec}$$
$$S_a = (0.5)\left(29 \frac{\text{ft}}{\text{sec}^2}\right) = 14.5 \text{ ft/sec}^2$$

(h) The base shear is given by Eq. 17.

$$V_{\text{horizontal}} = mS_a$$
$$= (35.4 \text{ slugs})\left(29 \frac{\text{ft}}{\text{sec}^2}\right)$$
$$= 1027 \text{ lbf}$$
$$V_{\text{vertical}} = (35.4 \text{ slugs})\left(14.5 \frac{\text{ft}}{\text{sec}^2}\right)$$
$$= 513 \text{ lbf}$$

(i) As determined in part (c), the assumptions result in a smaller period. Since lower values of

period give a higher acceleration (see Fig. 22), the assumptions result in the structure being designed for higher forces. Thus, the assumptions are conservative.

(j) Failing by bending, as opposed to fracture or collapse, is one of the indications of a ductile structure.

3. A two-story jail uses concrete shear walls as shown. None of the shear walls has openings, but only the east wall covers the entire length. The story height is 12 ft. The thickness of the roof, wall, and floor slabs are 5, 10, and 5 in, respectively. Assume all walls shown can be considered fixed piers.

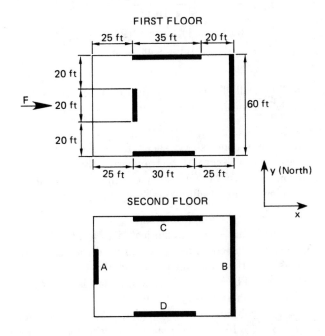

(a) What are the relative rigidities of the second story walls A, B, C, and D?

(b) Where is the center of rigidity?

(c) How would the shears in the second floor be determined? (Do not actually calculate the shears.)

(d) What is the effect on the first floor of offsetting wall A as shown?

Solution

(a) As stated in the problem, the walls are fixed piers. Use App. D to determine the rigidities of the fixed piers. The rigidities of perpendicular walls are taken as zero without regard to the h/d values.

wall	$\left(\dfrac{h}{d}\right)_{\text{E-W}}$	$\left(\dfrac{h}{d}\right)_{\text{N-S}}$	$R_{\text{E-W}}$	$R_{\text{N-S}}$
A	–	0.6	0	4.96
B	–	0.2	0	16.447
C	0.34	–	9.44	0
D	0.40	–	7.911	0

(b) The center of rigidity is located at (x_R, y_R). Distances are measured from the southwest corner.

$$x_R = \frac{(0\text{ ft})(4.96) + (80\text{ ft})(16.447)}{4.96 + 16.447} = 61.5$$

$$y_R = \frac{(0\text{ ft})(7.911) + (60\text{ ft})(9.44)}{7.911 + 9.44} = 32.6$$

(c) The wall shears are distributed in proportion to the second-floor wall rigidities.

(d) The second story shear from the outside (west) wall will have to be transferred through the second-story floor (first-story ceiling) slab to wall A below. Failure will probably occur in this slab.

4. The structure shown is subjected to a lateral loading of 0.3 g. The columns are set in hard rock concrete footings, but the tops can be considered to be pinned to the slab. The reinforced concrete slab is 6 in thick and has a finished density of 150 pcf (pounds per cubic foot). The columns are rectangular A36 steel tubing, TS $5 \times 5 \times (5/16$ in). Disregard the dead load of the columns, all live load, slab rotation, axial column compression, combined stresses, and column deflection. Determine if the columns are adequate.

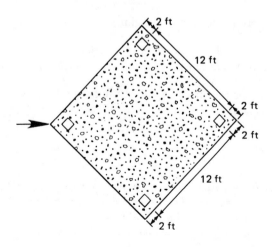

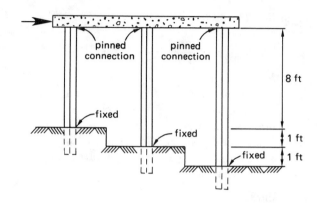

Solution

The properties of A500 Grade B TS $5 \times 5 \times$ (5/16 in) square structural tubing are found in the AISC manual. The modulus of elasticity is approximately 29×10^6 psi.

$$A = 5.52\text{ in}^2$$
$$I_x = I_y = 19.5\text{ in}^4$$

Although the tubing is oriented 45° from the plane of the principal axis, the moment of inertia per tube is still 19.5 in^4 in the direction of bending. This is because the tube is symmetrical and the product of inertia is zero.

Although the lower ends of the columns are fixed, the tops are pinned. Therefore, simple cantilever curvature occurs. From Table 7, the stiffness is

$$k = \frac{3EI}{L^3}$$

There are three different column lengths, so there are three stiffnesses.

$$k_{\text{8-ft}} = \frac{(3)\left(29 \times 10^6\ \dfrac{\text{lbf}}{\text{in}^2}\right)(19.5\text{ in}^4)}{(8\text{ ft})^3 \left(12\ \dfrac{\text{in}}{\text{ft}}\right)^3}$$

$$= \frac{1.7 \times 10^9\text{ in}^2\text{-lbf}}{8.847 \times 10^5\text{ in}^3} = 1922\text{ lbf/in}$$

$$k_{\text{9-ft}} = \frac{(2\text{ columns})(1.7 \times 10^9\text{ in}^2\text{-lbf})}{(9\text{ ft})^3 \left(12\ \dfrac{\text{in}}{\text{ft}}\right)^3}$$

$$= 2699\text{ lbf/in}$$

$$k_{\text{10-ft}} = \frac{1.7 \times 10^9\text{ in}^2\text{-lbf}}{(10\text{ ft})^3 \left(12\ \dfrac{\text{in}}{\text{ft}}\right)^3} = 984\text{ lbf/in}$$

The total stiffness consists of the sum of these three terms. (Notice that the stiffnesses for the two 9-ft columns were calculated as one sum.)

$$k_{\text{total}} = 1922 \frac{\text{lbf}}{\text{in}} + 2699 \frac{\text{lbf}}{\text{in}} + 984 \frac{\text{lbf}}{\text{in}}$$
$$= 5605 \text{ lbf/in}$$

The slab mass is

$$m = \frac{V\rho}{g_c} = \frac{(16 \text{ ft})(16 \text{ ft})(6 \text{ in})\left(150 \frac{\text{lbm}}{\text{ft}^3}\right)}{\left(12 \frac{\text{in}}{\text{ft}}\right)\left(32.2 \frac{\text{ft-lbm}}{\text{lbf-sec}^2}\right)}$$
$$= 596 \text{ slugs}$$

The seismic force is

$$F = ma = (596 \text{ slugs})(0.3 \text{ g})\left(32.2 \frac{\text{ft}}{\text{sec}^2\text{-gravity}}\right)$$
$$= 5760 \text{ lbf}$$

The resisting force in each tube is proportional to its relative rigidity (stiffness). The shortest tube has the highest stiffness, hence the shortest tube experiences the highest stress. The prorated portion of the force carried by the shortest tube is

$$F_{8\text{-ft}} = \left(\frac{1922 \frac{\text{lbf}}{\text{in}}}{5605 \frac{\text{lbf}}{\text{in}}}\right)(5760 \text{ lbf}) = 1975 \text{ lbf}$$

The moment at the base of the shortest tube is

$$M = FL = (1975 \text{ lbf})(8 \text{ ft})\left(12 \frac{\text{in}}{\text{ft}}\right)$$
$$= 1.896 \times 10^5 \text{ in-lbf}$$

The bending stress is

$$f_b = \frac{Mc}{I} = \frac{(1.896 \times 10^5 \text{ in-lbf})(\sqrt{2})\left(\frac{5 \text{ in}}{2}\right)}{19.5 \text{ in}^4}$$
$$= 34,400 \text{ lbf/in}^2 \text{ (psi)}$$

The minimum yield strength of normally-stocked A500 Grade B steel is 46 ksi. AISC Sec. F3.1 specifies the maximum allowable bending stress as 0.66 times the yield stress. Using the one-third increase in allowable stress from seismic loads, the maximum allowable seismic bending stress is 6 ksi.

$$F_b = (1.33)(0.66)F_y = (1.33)(0.66)(46,000 \text{ psi})$$
$$= 40,378 \text{ psi}$$

Since $f_b < F_b$, the design is acceptable.

5. An eight-story (including penthouse) office building is supported by a steel chevron-braced frame intended to carry all vertical loads and resist all seismic loads. There are no other load-supporting walls. The weights and heights of each floor are as given in the illustration. The building is located in seismic zone 4 on S_1 soil. Perform a seismic analysis consistent with the UBC, and determine (a) the base shear and (b) the seismic force at the third floor.

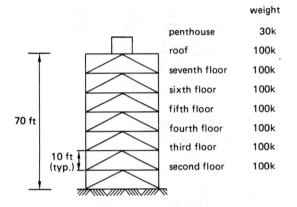

	weight
penthouse	30k
roof	100k
seventh floor	100k
sixth floor	100k
fifth floor	100k
fourth floor	100k
third floor	100k
second floor	100k

Solution

(a) From Table 11, $Z = 0.40$. From Table 12, $I = 1.00$. From Table 13, $S = 1.00$. From Table 14 for a concentric-braced steel frame, $R_w = 8$. Assuming the penthouse is not part of the structural system and is more "part and portion" than building, this is really a seven-story office building. From Eq. 64, the natural building period is

$$T = C_t(h_n)^{\frac{3}{4}} = (0.020)(70 \text{ ft})^{\frac{3}{4}} = 0.48 \text{ sec}$$

From Eq. 63,

$$C = \frac{1.25S}{T^{\frac{2}{3}}} = \frac{(1.25)(1.00)}{(0.48)^{\frac{2}{3}}} = 2.04$$

The building weight is

$$W = (7 \text{ stories})\left(100 \frac{\text{kips}}{\text{story}}\right) + 30 \text{ kips} = 730 \text{ kips}$$

From Eq. 62, the base shear is

$$V = \frac{ZICW}{R_w} = \frac{(0.40)(1.00)(2.04)(730 \text{ kips})}{8}$$
$$= 74.5 \text{ kips}$$

(b) From Eq. 69, the top force is zero since $T = 0.48\,\text{sec} < 0.7\,\text{sec}$.

Forming a table is the easiest way to determine the floor forces.

level x	h_x	W_x	$h_x W_x$
7	70	130	9100
6	60	100	6000
5	50	100	5000
4	40	100	4000
3	30	100	3000
2	20	100	2000
1	10	100	1000
		total	30,100

For the third floor (level 2),

$$\frac{h_x W_x}{\sum h_x W_x} = \frac{2000}{30,100} = 0.0665$$

From Eq. 70,

$$F_2 = \frac{(V - F_t)W_2 h_2}{\sum W_i h_i}$$
$$= (74.5\,\text{kips} - 0)(0.0665) = 4.95\,\text{kips}$$

6. The owner of a building is considering adding a wing to increase the size of the building. The existing floor plan and planned wing are shown in the figure below, along with structural walls and relative rigidities. All walls have the same thickness and height. The roof mass is to be disregarded. The owner is concerned about the lack of symmetry that the remodeled building will have. It has been determined that the base shear on the existing building is 90,000 pounds, parallel to the long (150-ft) dimension. An additional 15,000 pounds of base shear will be added by the new wing.

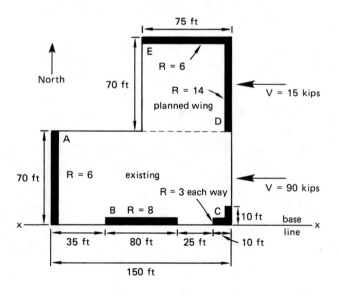

(a) Where will the center of mass be located when the wing is installed?

(b) Where will the center of rigidity be located when the wing is installed?

(c) What will be the torsional moment?

(d) Which active walls experience a negative torsional shear stress?

Solution

(a) The wall masses are proportional to their lengths since their thicknesses and heights are all the same. Using the 150-ft east-west wall as the baseline, for earthquakes in the east-west direction, the center of mass is located at a distance of

$$y_{\text{c.m.}} = \frac{\sum m_i y_i}{\sum m_i} =$$
$$\frac{(70)(35)+(80)(0)+(10)(0)+(10)(5)+(70)(70+35)+(75)(70+70)}{70+80+10+10+70+75}$$

$$= \frac{20,350}{315} = 64.6\,\text{ft (up from the baseline)}$$

(b) Walls running north-south do not contribute to rigidity for east-west earthquakes. Only walls B, C (part), and E are active. The center of rigidity is located at a distance of

$$y_{\text{c.r.}} = \frac{\sum R_i y_i}{\sum R_i}$$
$$= \frac{(8)(0) + (3)(0) + (6)(70 + 70)}{8 + 3 + 6}$$
$$= \frac{840}{17} = 49.4\,\text{ft (up from the baseline)}$$

(c) The actual eccentricity is

$$e_{\text{actual}} = y_{\text{c.m.}} - y_{\text{c.r.}} = 64.6\,\text{ft} - 49.4\,\text{ft} = 15.2\,\text{ft}$$

The accidental eccentricity required by the UBC is 5% of the transverse building direction.

$$e_{\text{accidental}} = (0.05)(70\,\text{ft} + 70\,\text{ft}) = 7\,\text{ft}$$

The total eccentricity is

$$e = e_{\text{actual}} + e_{\text{accidental}} = 15.2\,\text{ft} + 7\,\text{ft} = 22.2\,\text{ft}$$

The torsional moment is

$$M = Ve = (90\,\text{kips} + 15\,\text{kips})(22.2\,\text{ft})$$
$$= 2331\,\text{ft-kips}$$

(d) Walls B, C, and E resist the direct shear and are active. The base shear acts through the center of mass, tending to cause counterclockwise rotation about the center of rigidity. This is resisted by all walls (A, B, C, D, and E). However, walls B and C develop a clockwise stress, a direction opposite to the direct shear. Walls B and C develop a negative torsional shear stress.

7. A one-story, 55 ft by 70 ft masonry-walled building with a rigid diaphragm roof is constructed in the shape of a rectangle. One side of the building has two doors. There are no other openings. The rigidities of each wall are shown. Each wall is 10 in thick. Disregard the mass of the roof diaphragm.

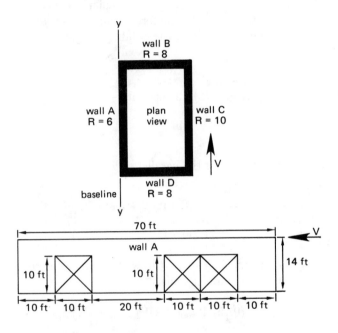

(a) What is the location of the center of rigidity?

(b) What is the torsional moment due to a total lateral force of 50,000 pounds?

(c) If 19,000 pounds of the lateral load are distributed to wall A, what loads are carried by each of the piers in wall A?

Solution

(a) For earthquakes in the direction shown, only walls A and C contribute to rigidity. The center of rigidity is located along a line parallel to baseline y-y and located a distance away of

$$x_{\text{c.r.}} = \frac{\sum R_i x_i}{\sum R_i} = \frac{(6)(0) + (10)(55)}{6 + 10} = \frac{550}{16}$$
$$= 34.4 \text{ ft}$$

(b) Although the openings could be disregarded, enough information is given to determine the location of the center of mass considering the openings. The mass of each wall is proportional to its area.

wall	area (mass)
A	$70 \times 14 - (3)(10)(10) = 680$
B	$55 \times 14 = 770$
C	$70 \times 14 = 980$
D	$55 \times 14 = 770$

The center of mass is located at

$$x_{\text{c.m.}} = \frac{\sum m_i x_i}{\sum m_i}$$

$$= \frac{(680)(0) + (770)\left(\frac{55}{2}\right) + (980)(55) + (770)\left(\frac{55}{2}\right)}{680 + 770 + 980 + 770}$$

$$= \frac{96,250}{3200} = 30.1 \text{ ft}$$

The actual eccentricity is

$$e_{\text{actual}} = x_{\text{c.r.}} - x_{\text{c.m.}} = 34.4 \text{ ft} - 30.1 \text{ ft} = 4.3 \text{ ft}$$

The accidental eccentricity is

$$e_{\text{accidental}} = (0.05)(55 \text{ ft}) = 2.8 \text{ ft}$$

The total eccentricity is

$$e = 4.3 \text{ ft} + 2.8 \text{ ft} = 7.1 \text{ ft}$$

The torsional moment is

$$M = Ve = (50,000 \text{ lbf})(7.1 \text{ ft}) = 355,000 \text{ ft-lbf}$$

(c) There are three piers in wall A. Due to the effect of the beam running along the top, assume the piers are fixed. The rigidities are found from App. D.

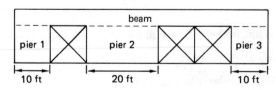

pier	h	d	h/d	R
1	10	10	1	2.5
2	10	20	0.5	6.154
3	10	10	1	2.5

The fraction of the total wall shear taken by piers 1 and 3 is

$$\frac{2.5}{2.5 + 6.154 + 2.5} = 0.22$$

The fraction taken by pier 2 is

$$1.00 - (2)(0.22) = 0.56$$

The pier shears are

$$V_1 = V_3 = (0.22)(19{,}000 \text{ lbf}) = 4180 \text{ lbf}$$
$$V_2 = (0.56)(19{,}000 \text{ lbf}) = 10{,}640 \text{ lbf}$$

8. A one-story building with masonry walls is constructed in a box shape. All walls are 24 ft high. The walls have a weight of 50 pounds per square foot (psf). The plywood roof and roof skylights have a weight of 25 psf. The roof carries a 20 psf live load of permanent air-conditioning equipment. The seismic load in the north-south direction is 15% of the participating building weight. Do not consider east-west earthquake performance.

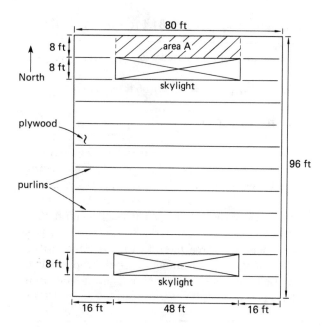

(a) What is the diaphragm force?

(b) What is the maximum wall shear force?

(c) What is the maximum wall chord force?

Solution

(a) For a north-south earthquake, the diaphragm force results from the acceleration of the roof mass and half of the short (80-ft) wall masses. The roof

weight (including the air-conditioning equipment weight) is

$$W_{\text{roof}} = \text{area} \times \text{loading}$$
$$= (80 \text{ ft})(96 \text{ ft})\left(20 \frac{\text{lbf}}{\text{ft}^2} + 25 \frac{\text{lbf}}{\text{ft}^2}\right)$$
$$= 345{,}600 \text{ lbf}$$

The weight of the perpendicular walls is

$$W_{\perp\text{walls}} = (80 \text{ ft})(24 \text{ ft})(2 \text{ walls})\left(50 \frac{\text{lbf}}{\text{ft}^2}\right)$$
$$= 192{,}000 \text{ lbf}$$

Only the top half of the perpendicular walls contributes to the diaphragm force. Since the seismic load is given as 15% of the building weight,

$$F_{\text{diaphragm}} = (0.15)(W_{\text{roof}} + \tfrac{1}{2}W_{\perp\text{walls}})$$
$$= (0.15)\left(345{,}600 \text{ lbf} + \frac{192{,}000 \text{ lbf}}{2}\right)$$
$$= 66{,}240 \text{ lbf}$$

(b) The weight of the parallel walls is

$$W_{\|\text{walls}} = (96 \text{ ft})(24 \text{ ft})(2 \text{ walls})\left(50 \frac{\text{lbf}}{\text{ft}^2}\right)$$
$$= 230{,}400 \text{ lbf}$$

Only the top half of all the walls contributes to shear in the parallel walls.

$$F_{\text{total}} = (0.15)(W_{\text{roof}} + \tfrac{1}{2}W_{\perp\text{walls}} + \tfrac{1}{2}W_{\|\text{walls}})$$
$$= (0.15)\left(345{,}600 \text{ lbf} + \frac{192{,}000 \text{ lbf}}{2}\right.$$
$$\left. + \frac{230{,}400 \text{ lbf}}{2}\right)$$
$$= 83{,}520 \text{ lbf}$$

Since there are two parallel walls, the maximum wall shear force is

$$V_{\|\text{wall}} = \frac{83{,}520 \text{ lbf}}{2 \text{ walls}} = 41{,}760 \text{ lbf/wall}$$

(c) From Eq. 92, the chord force in the perpendicular walls is

$$C = \frac{F_{\text{diaphragm}}L}{8b} = \frac{(66{,}240 \text{ lbf})(80 \text{ ft})}{(8)(96 \text{ ft})}$$
$$= 6900 \text{ lbf}$$

9. The plan view of a one-story, 12-ft-high masonry-walled retail shop is shown. Windows cover most of the front and half of one side. The building has a flexible plywood roof diaphragm with continuous roof struts (shown as lighter lines) criss-crossing and dividing the roof into subdiaphragms. Each strut is designed to serve as a chord, if necessary. The masonry shear walls have a dead weight of 55 pounds per square foot. The plywood roof has a dead weight of 15 pounds per square foot. For the purpose of the UBC base shear equation, $V = 0.183W$.

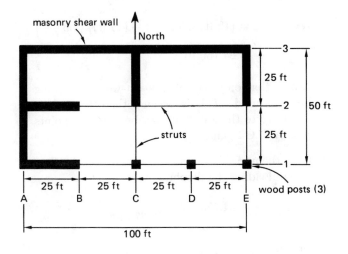

(a) What is the maximum diaphragm shear along lines A and E?

(b) What is the maximum chord or drag force at point C-1?

(c) What is the maximum drag force at point B-1 due to an east-west earthquake?

Solution

Notice that the earthquake direction is not given in this problem and must be considered variable.

(a) For north-south earthquakes, the roof is effectively divided into two subdiaphragms. One subdiaphragm is bounded by points (moving clockwise) A-1, A-3, C-3, and C-1. The other is bounded by points C-1, C-3, E-3, and E-1. Although each subdiaphragm is the same size, the accelerating masses are different.

For line A,

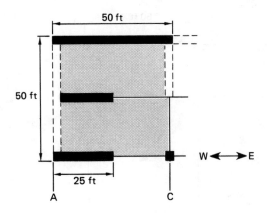

The weight of the roof is

$$W_{\text{roof}} = (50 \text{ ft})(50 \text{ ft})\left(15 \frac{\text{lbf}}{\text{ft}^2}\right) = 37{,}500 \text{ lbf}$$

The weight of the 50-ft wall is

$$W_{50 \text{ ft}} = (12 \text{ ft})(50 \text{ ft})\left(55 \frac{\text{lbf}}{\text{ft}^2}\right) = 33{,}000 \text{ lbf}$$

The weight of the two 25-ft walls totals 33,000 lbf also.

$$W_{25 \text{ ft}} = 33{,}000 \text{ lbf}$$

The seismic effect from the roof and the 50-ft wall is shared equally between walls A and C. Since the two remaining walls extend only halfway between A and C, a rational method of allocating their seismic effects to walls A and C must be used. Consider a simply supported beam loaded uniformly along the first half of its length. The reaction closest to the uniform load will carry $\frac{3}{4}$ of the total load. Therefore, assume that wall A carries $\frac{3}{4}$ of the seismic effect from the two short walls. (Other assumptions may be valid, depending on construction details.) Finally, assume that only the top half of the walls contribute to diaphragm shear.

The force in wall A is

$$F_A = 0.183W$$

$$= (0.183)\left[\frac{37{,}500 \text{ lbf}}{2} + \frac{\left(\frac{1}{2}\right)(33{,}000 \text{ lbf})}{2} + \frac{\left(\frac{3}{4}\right)(33{,}000 \text{ lbf})}{2}\right]$$

$$= 7206 \text{ lbf}$$

For line E,

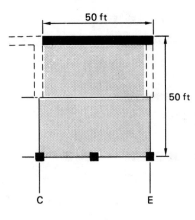

$$W_{\text{roof}} = 37{,}500 \text{ lbf}$$

$$W_{\text{E-W wall}} = (12 \text{ ft})(50 \text{ ft})\left(55 \frac{\text{lbf}}{\text{ft}^2}\right) = 33{,}000 \text{ lbf}$$

The effective east-west wall weight is

$$\frac{33{,}000 \text{ lbf}}{2} = 16{,}500 \text{ lbf}$$

$$
\begin{aligned}
F_{\text{diaphragm}} &= 0.183W \\
&= (0.183)(37{,}500 \text{ lbf} + 16{,}500 \text{ lbf}) \\
&= 9880 \text{ lbf}
\end{aligned}
$$

$$F_E = \frac{F_{\text{diaphragm}}}{2} = \frac{9880 \text{ lbf}}{2} = 4940 \text{ lbf}$$

(b) The earthquake direction is not given.

North-south earthquake

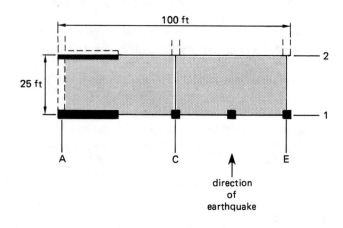

Drag force at point C-1

All diaphragm shear would frame into point C-2 since the wooden post at point C-1 can be assumed to have no lateral stiffness. Therefore, the maximum drag force at point C-1 due to a north-south earthquake is zero.

Chord force at point C-1

Half of the roof mass (corresponding to the tributary area bounded by points A-1, A-2, E-2, and E-1) and the upper half of the east-west walls in the tributary area contribute to the chord forces along lines 1 and 2. (Other interpretations might also be justified. For example, the entire roof might be considered.)

$$W_{\text{roof}} = (25 \text{ ft})(100 \text{ ft})\left(15 \frac{\text{lbf}}{\text{ft}^2}\right) = 37{,}500 \text{ lbf}$$

$$W_{\text{E-W walls}} = (12 \text{ ft})\left[25 \text{ ft} + \left(\tfrac{1}{2}\right)(25 \text{ ft})\right]\left(55 \frac{\text{lbf}}{\text{ft}^2}\right)$$

$$= 24{,}750 \text{ lbf}$$

Notice that only half of the east-west wall along line 2 is used. This is because the other half is tributary to the adjacent area bounded by points A-2, A-3, C-3, and C-2.

Only the upper half of the east-west walls is effective in loading the diaphragm.

$$\frac{24{,}750 \text{ lbf}}{2} = 12{,}375 \text{ lbf}$$

The diaphragm force is

$$
\begin{aligned}
F_{\text{diaphragm}} &= 0.183W \\
&= (0.183)(37{,}500 \text{ lbf} + 12{,}375 \text{ lbf}) \\
&= 9130 \text{ lbf}
\end{aligned}
$$

From Eq. 92, the chord force is

$$
\begin{aligned}
C &= \frac{F_{\text{diaphragm}}L}{8b} = \frac{(9130 \text{ lbf})(100 \text{ ft})}{(8)(25 \text{ ft})} \\
&= 4565 \text{ lbf}
\end{aligned}
$$

East-west earthquake

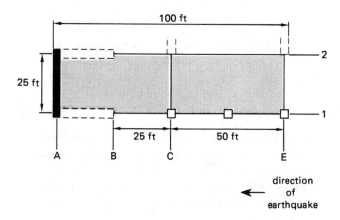

Drag force at point C-1

The collector along line 1 transfers half the diaphragm loading from the area bounded by points C-1, C-2, E-2, and E-1 into point B-1. (The other half is transferred to point B-2.) Since there are no north-south walls in this area, the only contribution to diaphragm force is the tributary roof weight. Between lines C and E, the roof weight is

$$W_{\text{roof}} = (25 \text{ ft})(50 \text{ ft}) \left(15 \, \frac{\text{lbf}}{\text{ft}} \right) = 18{,}750 \text{ lbf}$$

The subdiaphragm force is

$$F_{\text{subdiaphragm}} = 0.183W = (0.183)(18{,}750 \text{ lbf})$$
$$= 3430 \text{ lbf}$$

Half of this is transferred along the strut on line 1. At point C-1, the drag force is

$$D = \frac{3430 \text{ lbf}}{2} = 1715 \text{ lbf}$$

Chord force at point C-1

While there is a chord force along line C, the chord force at C-1 is zero because point C-1 is at the end of the chord.

(c) This is similar to part (b), except that the tributary roof area is 75 ft long instead of 50 ft long. Scaling up from the answer derived in part (b),

$$D = \left(\frac{75 \text{ ft}}{50 \text{ ft}} \right) (1715 \text{ lbf}) = 2570 \text{ lbf}$$

10. A 120-ft-wide by 240-ft-long warehouse in seismic zone 4 is oriented with its long dimension in the north-south direction. The walls are solid cast-in-place concrete, 10 ft high and 12 in thick, with no significant openings. The warehouse floor is loaded with 200 pounds per square foot. The roof consists of a 19/32-in, C-D plywood blocked diaphragm with an average weight of 15 pounds per square foot nailed to a 3-in (nominal) ledger with 10d nails. Consider only north-south earthquake motions.

(a) What diaphragm-to-perimeter edge nail spacing is required along the 240-ft wall?

(b) Sketch the method of interconnecting the diaphragm, ledger, and wall.

(c) Size the ledger bolts if they are spaced every 3 ft.

(d) What size grade 60 steel rebar should be used as the diaphragm chord?

(e) Where should the chord be located?

Solution

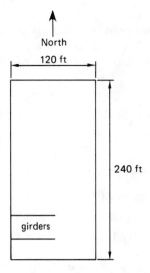

From Table 11, $Z = 0.40$. From Table 12, $I = 1.00$. Since the soil type is not known and the period of this box-type building will most certainly be low, use $C = 2.75$. From Table 14, $R_w = 6$.

From Eq. 62,

$$V = \frac{ZICW}{R_w} = \frac{(0.40)(1.00)(2.75)W}{6}$$
$$= 0.183W$$

(a) The roof weight is

$$W_{\text{roof}} = \frac{(120 \text{ ft})(240 \text{ ft}) \left(15 \, \frac{\text{lbf}}{\text{ft}^2} \right)}{1000 \, \frac{\text{lbf}}{\text{kip}}} = 432 \text{ kips}$$

The east-west walls contribute to diaphragm loading. Concrete has a weight density of approximately 150 pounds per cubic foot.

$$W_{\text{E-W walls}} = \frac{(10 \text{ ft})(2 \text{ walls})(120 \text{ ft})(12 \text{ in}) \left(150 \, \frac{\text{lbf}}{\text{ft}^3} \right)}{\left(12 \, \frac{\text{in}}{\text{ft}} \right) \left(1000 \, \frac{\text{lbf}}{\text{kip}} \right)}$$
$$= 360 \text{ kips}$$

Only the top half of the east-west walls are effective in loading the diaphragm.

$$\frac{360 \text{ kips}}{2} = 180 \text{ kips}$$

Since this is a warehouse, a minimum of 25% of the live load must be added to the building weight. (See Sec. 97. It could also be argued that the storage sits on grade and does not add to the inertial mass.)

$$W_{\text{live}} = (0.25)(120 \text{ ft})(240 \text{ ft})\left(200 \frac{\text{lbf}}{\text{ft}^2}\right)$$
$$= 1.44 \times 10^6 \text{ lbf } (1440 \text{ kips})$$

The diaphragm force is

$$F_{\text{diaphragm}} = 0.183W$$
$$= (0.183)(432 \text{ kips} + 180 \text{ kips} + 1440 \text{ kips})$$
$$= 376 \text{ kips}$$

The diaphragm force is resisted by connections along two walls, each of 240 ft length. The shear force per unit length along the north-south walls is

$$V = \frac{(376 \text{ kips})\left(1000 \frac{\text{lbf}}{\text{kip}}\right)}{(2 \text{ walls})\left(240 \frac{\text{ft}}{\text{wall}}\right)} = 783 \text{ lbf/ft}$$

From Table 17, 2-in nail spacing provides 820 lbf/ft of shear resistance.

(b)

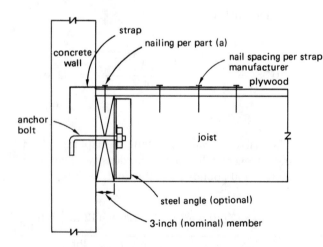

(c) Use Table 21. (See Sec. 148 B.) The ledger is a nominal 3 in thick. Doubling this gives 6 in. Since the actual thickness is less than the nominal thickness, use the $5\frac{1}{2}$-in row. Use the double shear values. The shear loading is parallel to the grain. Since the bolts are spaced every 3 ft, the shear load is

$$V = \left(3 \frac{\text{ft}}{\text{bolt}}\right)\left(783 \frac{\text{lbf}}{\text{ft}}\right) \approx 2350 \text{ lbf/bolt}$$

Doubling this in order to use the table in a concrete-wood connection, and multiplying by $\frac{3}{4}$ (the reciprocal of $\frac{4}{3}$) to reduce the seismic load to a normal duration load,

$$V = 2 \times 2350 \frac{\text{lbf}}{\text{bolt}} \times \frac{3}{4} = 3525 \text{ lbf/bolt}$$

Use a $\frac{7}{8}$-in diameter bolt that has a table strength of 3900 lbf/bolt.

(d) From Eq. 92, the maximum chord force in the short walls will be

$$C = \frac{F_{\text{diaphragm}}L}{8b}$$
$$= \frac{(376 \text{ kips})\left(1000 \frac{\text{lbf}}{\text{kip}}\right)(120 \text{ ft})}{(8)(240 \text{ ft})}$$
$$= 23{,}500 \text{ lbf}$$

Since the allowable stress for grade 60 rebar is 24,000 psi (see Sec. 132), the required bar size is given by Eq. 95.

$$A = \frac{C}{1.33 \times \text{allowable tensile stress}}$$
$$= \frac{23{,}500 \text{ lbf}}{(1.33)\left(24{,}000 \frac{\text{lbf}}{\text{in}^2}\right)} = 0.736 \text{ in}^2$$

Use a no. 8 bar (diameter, 1 in; area, 0.78 in^2).

(e) The chord should be located in the plane of the diaphragm. See Fig. 70.

QUALITATIVE PROBLEMS

1. List the numbers corresponding to the types of individuals who have the legal authority in California to perform the functions listed. (Unless noted otherwise, references in parentheses are to the California Business and Professions Code, Chapters 7 (Professional Engineers) and 15 (Professional Land Surveyors).)

1) any licensed civil engineer

2) civil engineer licensed before January 1, 1982

3) civil engineer licensed after January 1, 1982

4) licensed civil engineer specializing in structures

5) licensed (California) soils engineer

6) licensed structural engineer

7) engineer licensed in any field

8) unlicensed civil engineer specializing in structures, under the responsible charge of a licensed civil engineer

9) any unlicensed civil engineer

10) licensed architect

11) licensed land surveyor

12) licensed photogrammetric surveyor

13) licensed contractor

14) licensed building designer

15) any member of the general public

16) no one

(a) use the title "consulting engineer" publicly

(b) use the title "civil engineer" publicly

(c) use the title "structural engineer" publicly

(d) use the title "soils engineer" publicly

(e) use the title "land surveyor" publicly

(f) personally perform civil engineering work

(g) solicit civil engineering work for others

(h) perform architectural work

(i) sign, stamp, and seal civil engineering design plans developed by an unlicensed engineering subordinate under the individual's direct engineering control

(j) sign, stamp, and seal civil engineering design plans developed by an unlicensed engineer in the individual's company whose paycheck the individual signs

(k) sign, stamp, and seal civil engineering design plans developed by a qualified, unlicensed, moonlighting engineer who pays the individual

(l) sign, stamp, and seal civil engineering design plans developed by a qualified, unlicensed, engineer who does not pay the individual

(m) allow a qualified, unlicensed person to use a registered civil engineer's stamp, seal, or registration number

(n) design a single-story, wood-framed residence

(o) design a two-story, wood-framed building

(p) design a five-story, concrete-framed building

(q) design a five-story, steel-framed building

(r) design an above-ground water tower structure

(s) design a hospital building

(t) design a new public school building

(u) design a new private school building

(v) design a steel bridge

(w) inspect an earthquake-damaged building within 30 days of the event, without payment, when requested by the local building official

(x) supervise the construction of designed structures

(y) perform land surveying work for hire

(z) solicit land surveying work for others

(aa) perform a survey of public lands

(bb) perform a survey of private lands to be subdivided

(cc) lay out a construction site using surveying knowledge, methods, and equipment

(dd) gather in the field information to be placed on a deed or record-of-survey map

(ee) file a record-of-survey map with the county

(ff) administer oaths, certify oaths, and take testimony under oath to identify lost corners

Answers

(a) 1, 5, 6, 7, 12 (6704, 6732)

(b) 1, 5, 6 (6704, 6732, 6734)

(c) 6 (6703, 6704, 6732, 6736)

(d) 5 (6704, 6732, 6736.1, 6763)

(e) 11 (6731, 8708, 8725, 8731)

(f) 1, 5, 6 (6731.2, 8726.1)

(g) 15

(h) 10 (6737)

(i) 1, 5, 6 (6730.2, 6735, 6740)

(j) 16 (6703, 6735)

(k) 16 (6703, 6735)

(l) 16 (6703, 6735)

(m) 16 (6732, 6735)

(n) 15 (6737, 6737.1)

(o) 15 (6737, 6737.1)

(p) 1, 5, 6, 8, 10 (6737)

(q) 1, 5, 6, 8, 10 (6737)

(r) 1, 5, 6, 8, 10 (6737)

(s) 6 (Refer to Health and Safety Code, Div. 12.5, Chap. 1 Hospitals, Sec. 15048)

(t) 6

(u) 1, 5, 6, 8, 10 (6737)

(v) 1, 5, 6, 8

(w) 1, 5, 6, 7 (6706)

(x) 1, 5, 6, 10, 13 (6731, 6731.3, 6735.1)

(y) 2, 11, 12 (6731.2, 8726.1, 8731, 8775)

(z) 15

(aa) 2, 11, 12 (8708, 8731, 8775)

(bb) 2, 11, 12 (8708, 8726, 8731, 8775)

(cc) 12, 15

(dd) 2, 11, 12 (8726, 8731, 8775)

(ee) 2, 11, 12 (8731, 8762, 8775)

(ff) 1, 2, 5, 6, 11, 12 (8760, 8775)

2. How does the UBC cover the design of bridges?

Answer

The UBC covers only the design of buildings and some other building-like structures. Design of bridges is not covered in the UBC. This subject is covered in CALTRANS and American Association of State and Highway Transportation Officials (AASHTO) publications. There are no official standards, however, for the retrofit of bridges.

3. What provisions does the UBC make for buildings subject to landslides, liquefaction, subsidence, gross differential settlement, or for those built close to a major ground-breaking fault?

Answer

None. The UBC provides rules for the design of buildings that will resist typical ground shaking. The UBC assumes the engineer will use good judgment in avoiding inherently dangerous locations.

4. What type of information will generally be supplied by the geotechnical engineer working on a building design team?

Answer

The geotechnical engineer will determine the (a) type of soil (i.e., sand, clay, or rock); (b) depth of water table; (c) depth to bedrock; (d) proximity to a fault; (e) maximum credible earthquake; and (f) likelihood of liquefaction, slides, subsidence, and differential settlement.

5. Which structural system resists lateral loads by flexure in members and joints?

Answer

Only moment-resisting frames resist lateral loads in this manner.

6. What is the meaning of the term "secondary stress" as it relates to a moment-resisting frame?

Answer

Primary stresses are the compressive and tensile forces that act uniformly on the cross section of the member. Secondary stresses are bending stresses that result from distortion of the frame when resisting lateral loads by flexure.

7. (a) What structural elements transfer lateral loads to vertical elements? (b) What structural elements transfer lateral loads to lower levels and the foundation?

Answer

(a) Horizontal elements such as diaphragms, horizontal bracing, and beams in moment-resisting frames transfer horizontal loads to vertical elements. (b) Vertical elements such as shear walls, braced frames, and columns in moment-resisting frames transfer lateral loads to lower levels.

8. What is the meaning of the term "weak (soft) story"? Give an example.

Answer

A soft story does not have as much lateral force resistance as the stories above. An example is a moment-resisting frame supported by long columns over an open plaza below.

9. What restrictions does the UBC place on situations where the type of structural system is different for different levels of a multistory building? What are the exceptions?

Answer

The value of R_w used in the design of one level must be less than or equal to the value of R_w used to design the level above. An exception is where the story above constitutes less than 10% of the total structure weight (i.e., is very light) [UBC Sec. 2334(c)2].

10. What is the meaning of the term "irregular building"? Give two examples.

Answer

For the purpose of the UBC, an irregular building meets one or more of the characteristics in UBC Table 23-M. Examples are (1) a three-story, L-shaped building and (2) a five-story, square building with an open plaza comprising 60% of the floor area on level 3.

11. What is the meaning of the word "pounding"?

Answer

Pounding refers to adjacent buildings coming into contact with each other. One building can sway into another and pound it. The danger is greater when floor slabs of one building pound the columns of another; the danger is less when the floor slabs are at the same elevation. Up to 20% of the building failures in the 1985 Mexico City earthquake are thought to have been caused by pounding. Some of the buildings damaged in the 1989 Loma Prieta earthquake in the Watsonville-Santa Cruz area were only 6 in apart and were damaged because they pounded each other.

12. When can the UBC's dynamic analysis method be used to determine the seismic force on a building? When can it not?

Answer

The dynamic method described by the UBC [Sec. 2335] can always be used. It is the static method that is limited and that must satisfy certain conditions [UBC Sec. 2333(h)].

13. What is the maximum span-to-width ratio for a plywood roof diaphragm?

Answer

The maximum span-to-width ratio for a roof diaphragm is 4:1 [UBC Table 25-I].

14. It is generally stated and understood that flexible diaphragms cannot transmit torsional shear stress to vertical resisting elements. Is this true for a flexible diaphragm that is cantilevered off of a vertical wall?

Answer

This is a tricky question. Any eccentric mass can cause torsion. A cantilevered flexible diaphragm, when acted upon by a seismic force perpendicular to its cantilevered dimension, will cause the wall to twist. However, this is different than transmitting torsion caused by one component to another. A cantilever wall can cause torsion; it cannot transmit torsion.

15. It is generally stated and understood that the lateral loads resisted by vertical elements attached to rigid diaphragms are proportional to the element rigidities, and the lateral loads resisted by vertical elements attached to flexible diaphragms are proportional to tributary areas. How are lateral loads resisted by closely-placed vertical elements that are arranged in-line, are parallel to an earthquake's motion, and are attached to a single flexible diaphragm?

Answer

Since all of the elements have the same tributary area, they will resist the lateral load in proportion to their relative rigidities.

16. Which is more life-threatening: shear cracking in a seismically-detailed concrete column or flexural cracking of a seismically-detailed concrete shear wall?

Answer

Cracking in a shear wall is probably more serious than cracking in a column. A properly-detailed column should not lose its load-bearing capacity merely because of cracking. The strict seismic detailing is intended to keep concrete in a column intact and confined even if it cracks. However, such confinement is not as complete in shear walls.

17. A tank on the roof of a building contains hazardous chemicals.

(a) What importance value, I, should be used in calculating the seismic force on the building?

(b) What importance value, I, should be used in calculating the seismic force on the tank?

Answers

   (a) 1.25 [UBC Table 23-L]

   (b) 1.50 [UBC Sec. 2336(b) Exception 2]

18. What is the basic distinction between ordinary and special moment-resisting frames?

Answer

   A special moment-resisting frame has been carefully detailed to remain ductile. An ordinary moment-resisting frame does not have this detailing.

19. Two buildings have the same mass, but one building has a shorter natural period than the other building. All other factors being equal, which building will experience the larger seismic force?

Answer

   Most response spectra show that the lower the natural period, the higher the acceleration experienced by the building. Therefore, the building with the shorter period will probably experience the larger seismic force.

20. In UBC seismic zone 4, which material is most likely to be less expensive when building a 30-story moment-resisting frame: steel or concrete?

Answer

   This is a controversial question whose answer may depend on loyalties. However, more high-rise buildings seem to be built out of steel than out of concrete. All things being equal, steel is probably less expensive.

21. What consideration should be given to the design of a building that resists lateral force by a combination of braced frame and shear wall action?

Answer

   Braced frames and shear walls have different stiffnesses and may deflect different amounts. This will cause a separation where the two resisting systems meet. The resisting elements must be proportioned so that the deflections are equal for both resisting systems.

22. In an extreme earthquake, what type of fascia would sustain the most damage: glass or concrete?

Answer

   This is a fairly vague question since only the type of fascia material (and not the mounting method) is indicated. Glass has no ductility, so glass fascia probably would not fare well in an extreme earthquake. Concrete fascia would probably have been cast with continuous bar or mesh reinforcing. This reinforcing would help the concrete fascia remain intact when flexed.

23. For small buildings with only one or two floors, which of the different structural systems are more cost-effective? (Limit your discussion to plywood shear wall construction, masonry wall systems, steel braced frames, stiff-redundant steel systems, concrete moment-resisting frames, steel moment-resisting frames, and dual systems.)

Answer

   Small buildings with only one or two floors can be built using any of the structural systems listed, although dual, redundant, and moment-resisting frame systems probably would not be used. The systems in order of increasing cost are

   1. plywood shear wall construction

   2. masonry wall systems

   3. steel braced frames

   4. dual systems

   5. stiff-redundant systems

   6. concrete moment-resisting frames

   7. steel moment-resisting frames

24. For tall buildings with more than ten floors, which of the different structural systems are more cost-effective? (Limit your discussion to plywood shear wall construction, masonry shear wall (box) systems, steel braced frames, stiff-redundant steel systems, concrete moment-resisting frames, steel moment-resisting frames, and dual systems.)

Answer

   Plywood, masonry, and dual systems would not be used for a building with ten floors. Exceptionally tall buildings must be built either exceptionally stiff (e.g., the Empire State Building) or must use moment-resisting frames. Most modern tall buildings in California are constructed of steel. The logical conclusion is that these are less expensive than concrete buildings. Stiffness achieved at the expense of multiple redundancy is the most expensive. Though necessary in the early history of tall building construction, designing stiffness through redundancy is no longer practiced.

25. The floors in the top half of a tall multistory building are much smaller (in plan view) than the floors in the bottom half of the building.

(a) What are the problems associated with this design?

(b) How would you counteract the problems?

Answer

(a) This question is essentially about the subject of setbacks. The main problem is that the upper half would have a different period and different mode shape than the lower half. The upper floors could oscillate out-of-phase with the upper floors. This is referred to as "whipping action." Large stresses would be generated when the two sections were 180° out-of-phase. The stress would be most severe at the setback points.

(b) The upper half of the building must be designed so that, though smaller, it is as stiff or flexible as the lower half. There are many ways of increasing stiffness, including adding bracing, changing the spacing or number of interior members, and increasing member sizes. (Since the mass of the upper stories is reduced, just keeping the column and beam sizes the same as in the lower stories would help.) In some cases, a different construction material could be used. It is not generally practical to add stiffness by starting new columns at an upper floor.

26. An air conditioning unit is placed on a plywood roof diaphragm. What effect does the new unit have on the damping ratio of the roof?

Answer

None. The damping ratio of the roof is a function of the roof material, design, and quality of construction.

27. Draw simple diagrams that show how a three-story frame would fail in (a) beam-hinge mode, (b) column-hinge mode, and (c) soft-story (also known as "weak-story") mode. Show all plastic hinge points.

Answer

Plastic hinges are shown as solid bullets.

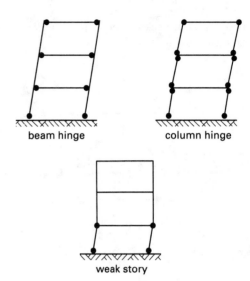

beam hinge   column hinge

weak story

28. Draw simple diagrams that show how a four-story frame would fail in (a) weak-column mode (i.e., when the beams were stronger than the columns) and (b) weak-beam mode (i.e., when the columns were stronger than the beams). Show all plastic hinge points.

Answer

Plastic hinges are shown as solid bullets.

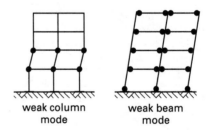

weak column   weak beam
mode   mode

29. Draw the connections necessary to anchor the floor diaphragms shown to the side of a CMU (concrete masonry unit) wall. Show and label all connectors and other elements. Describe the load path (i.e., how the load is transferred from horizontal members to the vertical members). No calculations are necessary and no specific spacings need to be specified. Assume positive attachment to the wall is spaced approximately every 4 ft.

(a) plywood floor on 2× joists attached to 4 × 10 ledger

(b) plywood supported directly by 4 × 10 ledger

(c) corrugated steel decking supported by steel ledger angle

(d) steel plate supported by steel ledger angle

(e) poured gypsum deck on metal form deck supported on steel ledger angle

Answers

(a)

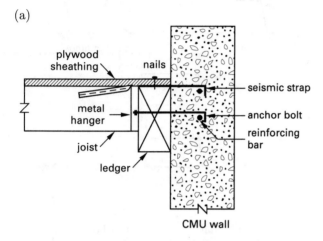

(b)

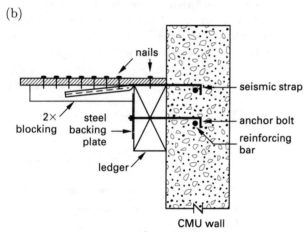

(c)

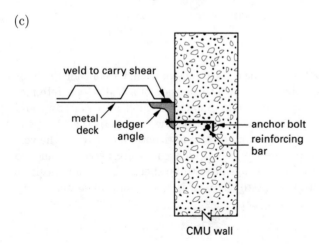

(d)

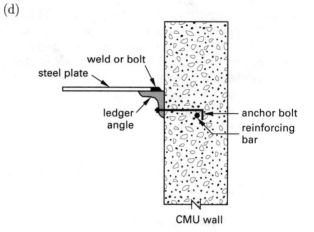

(e)

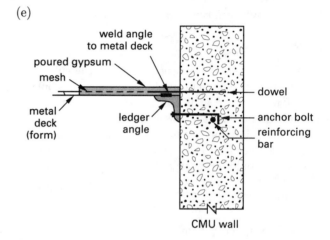

30. A plywood diaphragm is supported by 2× joists. The joists are supported by a wood ledger attached to a CMU (concrete masonry unit) wall. Detail the connection between a joist and the ledger if the joist does not coincide with a wall anchor bolt or seismic strap. Assume the ledger strength is adequate.

Answer

This problem is somewhat contrived because if the application is really critical, the joint should be connected directly to the wall. The intent of this problem is to design a positive connection between the joist and ledger. In doing so, it must be recognized that (a) provisions must be made to avoid tension splits in the joist, and (b) connector pull-out strengths must be considered in attaching the joist to the ledger. Toe-nailing is obviously inadequate. If the joist is attached to the ledger with a commercial hanger having a row of vertical nails, the transmitted force will be limited by edge distance. The detail shown uses (a) nailing or bolting along the joist to avoid tension tear-out and (b) lag bolting to avoid connector pull-out.

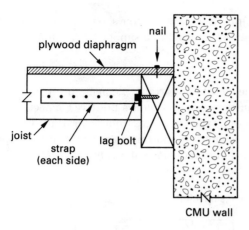

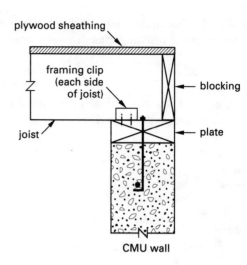

31. Detail a connection for a 2× joist supporting a plywood floor diaphragm sitting directly on a wall plate on top of a CMU (concrete masonry unit) wall. How does your connection avoid cross-grain tension in the wall plate?

Answer

The connection shown avoids cross-grain tension in the wall plate by avoiding any connection to the wall plate. Lateral loads are transmitted in bearing through the anchor bolt. The wall plate remains in vertical compression at all times.

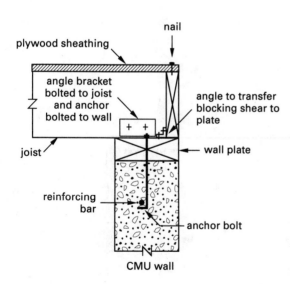

32. Describe how the connection between the joist and masonry unit wall may fail in cross-grain tension. How could you retain the basic design and eliminate the cross-grain tension?

Answer

If the lateral load is from left-to-right, the plate will be placed in compression by the anchor bolt reaction acting to the left. However, if the lateral load is from right-to-left, the plate will be placed in cross-grain tension by the anchor bolt reaction acting to the right. This design can be "fixed" by adding a second framing clip to the right of the anchor bolt.

33. Illustrate how plywood shear walls directly above each other on two different levels could be interconnected. (Exterior sheathing cannot be used to interconnect the walls.)

Answer

The most common method is to use a connecting tierod (tie down), as shown.

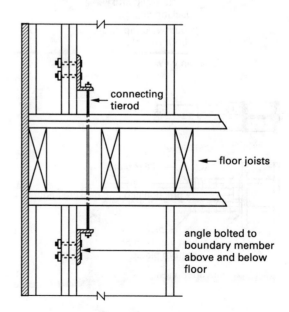

34. Draw and detail three types of column-girder joints for special moment-resisting frames constructed from steel.

Answer

The three methods shown are (a) a butt-welded joint, (b) a fillet-welded joint, and (c) a bolted joint.

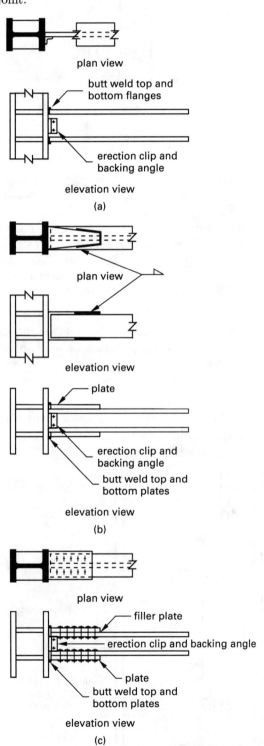

35. Draw a typical section showing how the end of a bridge section would be supported by an abutment.

Answer

The most important element in a bridge-to-abutment connection is a positive connection that prevents the two pieces from separating. Elements of secondary importance are the bearing, expansion joint, and shock-absorbing element.

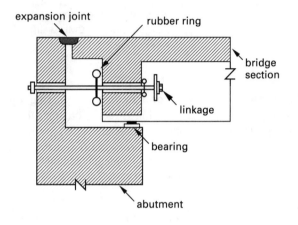

APPENDICES

APPENDIX A
Conversion Factors

Multiply	By	To Obtain	Multiply	By	To Obtain
acre	43,560	ft^2	kg	2.20462	lbm
BTU	778.17	ft-lbf	kg	0.06852	slug
BTU	1.055	kJ	kip	1000	lbf
BTU/h	0.293	W	kJ	0.9478	BTU
BTU/lbm	2.326	kJ/kg	kJ	737.56	ft-lbf
BTU/lbm-°R	4.1868	kJ/kg·K	kJ/kg	0.42992	BTU/lbm
cm	0.3937	in	kJ/kg·K	0.23885	BTU/lbm-°R
cm^3	0.061024	in^3	km	3280.8	ft
erg	7.376×10^{-8}	ft-lbf	km	0.6214	mi
ft	0.3048	m	km/h	0.62137	mi/h
ft^3	7.481	gal	kPa	0.14504	lbf/in^2
ft^3	0.028317	m^3	kW	737.6	ft-lbf/sec
ft-lbf	1.356×10^7	erg	kW	1.341	hp
ft-lbf	1.35582	J	l	0.03531	ft^3
ft/sec^2	0.0316	gravities	l	0.001	m^3
in/sec^2	0.002591	gravities	lbf	4.4482	N
gal	0.13368	ft^3	lbf/ft^2	144	lbf/in^2
gal	3.7854×10^{-3}	m^3	lbf/in^2	6894.8	Pa
gal/min	0.002228	ft^3/sec	lbm	0.4536	kg
g/cm^3	1000	kg/m^3	lbm/ft^3	0.016018	g/cm^3
g/cm^3	62.428	lbm/ft^3	lbm/ft^3	16.018	kg/m^3
gravities	32.2	ft/sec^2	m	3.28083	ft
gravities	386	in/sec^2	m^3	35.3147	ft^3
gravities	9.81	m/s^2	mm	0.03937	in
hp	2545	BTU/h	m/s^2	0.1019	gravities
hp	33,000	ft-lbf/min	mi	1.609	km
hp	550	ft-lbf/sec	mi/h	1.6093	km/h
hp	0.7457	kW	N	0.22481	lbf
in	2.54	cm	Pa	1.4504×10^{-4}	lbf/in^2
in	25.4	mm	slug	32.174	lbm
in^3	16.387	cm^3	W	3.413	BTU/h
J	0.73756	ft-lbf			

APPENDIX B
Definitions

Base: The level at which the earthquake motions are imparted to the structure.

Box system: A structural system without a complete vertical load-carrying frame. In this system, the lateral forces are resisted by shear walls or braced frames.

Braced frame: A truss system that is provided to resist lateral forces and in which the members are subjected primarily to axial stresses.

Confined concrete: Concrete confined by closely-spaced ties restraining it in a direction perpendicular to the applied stress.

Critical damping: The amount of damping that results in the system recovering from an initial deflection in the minimum amount of time, without an amplitude reversal.

Damping: The characteristic that reduces the vibrational energy, primarily by friction.

Dip: The angle that a stratum or fault makes with the horizontal.

Dip slip: The component of the slip parallel with the dip of the fault.

Ductile moment-resisting frame: A frame with rigid connections between columns and girders that is ductile at potential yielding points. See also *Moment-resisting frame*.

Elastic rebound theory: A seismic theory, based on the tectonic plate concept, which proposes that stresses are created in fault lines by shifting of the tectonic plates, and that faults resist motion until the accumulated stress overcomes the internal friction.

Equivalent static load: A single horizontal load, as defined in the UBC, for which an earthquake-resistant building should be designed.

Essential facility: A facility that must remain functional after a major earthquake.

Fault: A fracture or fracture zone along which the sides can move relative to one another and parallel to the fracture.

Fault creep: Continuous displacement along a fault at a slow and varying rate that is usually not accompanied by noticeable earthquakes.

Fault displacement: Relative movement of the two sides of a fault, measured in any specified direction (usually parallel to the fault).

Fault gouge: Filler material that forms between two plates sliding against each other.

Fault sag: A narrow tectonic (generally earth-filled) depression common in strike-slip fault zones, less than a few hundred feet wide and approximately parallel to the fault zone. See also *Sag pond*.

Fault scarp: A cliff or steep slope formed by displacement of the ground surface.

Fracture: A general term for a break, joint, or fault in the Earth.

Frame: A two-dimensional structural system without bearing walls that is composed of interconnected laterally-supported members, and that functions as a self-contained unit.

Gouge: See *Fault gouge*.

Graben (plural, *graben*): A fault block, generally long and narrow, that has dropped down relative to the adjacent blocks.

Hoop: A one-piece closed tie or continuously wound tie that encloses longitudinal reinforcement.

Hypocenter: The actual location of the earthquake beneath the Earth's center.

Igneous rock: Rocks formed by the solidification of molten magma.

Lateral force-resisting system: The part of the building that resists earthquake and wind forces.

Left-normal slip: Fault displacement consisting of nearly equal components of left and normal slips.

Left slip: Strike-slip displacement in which the block across the fault moves to the left.

Liquefaction: The loss of load-carrying ability in loose, usually saturated, soil or sand.

Moment-resisting frame: A vertical load-carrying frame in which the members and joints are capable of resisting forces primarily by flexure.

Normal fault: Any fault (including those with vertical slip) in which the block above an inclined surface moves downward relative to the block below the fault surface.

Normal slip: Vertical displacement of a fault.

Oblique slip: A combination of strike slip and reverse slip.

APPENDIX B (continued)
Definitions

Parapet: A low wall or railing (including decorative panels) extending, usually vertically, above the roof line.

Plastic hinge: A region where the yield moment strength of a flexural member is exceeded and that experiences significant rotation.

Reserve energy: Energy that a ductile system is capable of absorbing in the inelastic region.

Resonance: A condition existing when the frequency of excitation is the same as the natural frequency of the building or soil.

Reverse fault: A fault in which the block above an inclined fault surface moves upward relative to the block below the fault surface.

Right-normal slip: Fault displacement consisting of nearly equal components of right and normal slips.

Right slip: Strike-slip displacement in which the block across the fault from an observer moves to the right.

Rigid frame: A vertical load-carrying frame in which the members and joints resist forces by rotation and flexure. See also *Moment-resisting space frame*.

Sag pond: A fault sag that has filled with water.

Shear wall: A wall designed to resist lateral forces parallel to the plane of the wall.

Slip: The relative displacement, measured on the surface, of two points on opposite sides of a fault.

Space frame: A three-dimensional structural system without bearing walls that is composed of interconnected laterally supported members and that functions as a self-contained unit.

Special ductile frame: A structural frame designed to remain vertically functional after the formation of plastic hinges from reversed lateral displacements.

Special shear wall: A reinforced concrete shear wall designed and detailed in accordance with the special UBC provisions.

Stirrup tie: A closed stirrup completely encircling the longitudinal members of a beam or column and conforming to the definition of a hoop.

Strike: The horizontal direction or bearing of the fault on the surface.

Strike-slip: The horizontal component of slip, parallel to the strike of the fault.

Strike-slip fault: A fault in which the slip is approximately in the direction of the strike.

Supplementary crosstie: A tie with a standard 180-degree hook at each end.

Tectonic: Pertaining to or designating the internal and external rock structures and features caused from crustal and subcrustal activity deep in the Earth.

Tectonic creep: Fault creep of tectonic origin.

Transcurrent fault: See *Strike-slip fault*.

Wrench fault: See *Strike-slip fault*.

APPENDIX C
Chronology of Important California Earthquakes

Date	Fault	Richter Magnitude	Surface Effects and Significance
1836	Hayward	7.0 (est.)	Ground breakage
1838	San Andreas	7.0 (est.)	Ground breakage
1852	Big Pine		Possible ground breakage
1857	San Andreas	8.0 (est.)	Right-lateral slip, possibly as much as 30 ft
1861	Calaveras		Ground breakage
1868	San Andreas		Long fissure in earth at Dos Palmas
1868	Hayward	7.0 (est.)	Strike-slip
1872	Owens Valley zone	8.3 (est.)	Right-lateral slip of 16–20 ft. Left-lateral slip may also have occurred. Vertical slip, down to east, of 23 ft.
1890	San Andreas		Fissures in fault zone; railroad tracks and bridge displaced
1899	San Jacinto	6.6 (est.)	Possible surface evidence
1906	San Andreas	8.3	Known as the "San Francisco earthquake." Right-lateral slip up to 21 ft. Resulted in the formation of the California State Earthquake Investigation Commission.
1922	San Andreas	6.5	Ground breakage
1925	Mesa/Santa Ynez	6.3	Known as the "Santa Barbara earthquake of 1925." U.S. Coast and Geodetic Society was directed to study the field of seismology.
1927	Santa Ynez	7.5	Occurred offshore in a submarine trench and was felt on land
1933	Newport-Inglewood	6.3	Known as the "Long Beach earthquake of 1933." Extensive property damage and loss of life. Many school buildings were destroyed. Resulted in the passage of the Field Act. The Division of Architecture of the State Department of Public Works was assigned responsibility to approve new buildings used for schools. The Riley Act was also passed, which set minimum requirements for lateral force design.
1934	San Andreas	6.0	Ground breakage
1934	San Jacinto	7.1	Ground breakage
1940	Imperial	7.1	Known as the "El Centro earthquake." 40 mi of surface faulting. 80% of Imperial buildings were damaged. However, no Field Act school buildings were damaged. This was the first major earthquake to yield accelerograph data on building periods. A maximum acceleration (ground) of 0.33 g was experienced.
1947	Manix	6.4	Left-lateral slip of 3 in
1950	(unnamed)	5.6	Vertical slip, down to west, of 5–8 in along the west edge of Fort Sage Mountains
1951	Superstition Hills	5.6	Slight right-lateral slip
1952	White Wolf	7.7	Known as the "Kern County earthquake," and the "Arvin-Tehachapi earthquake." Extensive building damage to old buildings. Little or none to properly designed and Field Act buildings. Confirmed the requirement for proper design.
1956	San Miguel	6.8	Right-lateral slip, 3 ft; vertical slip, down to southwest, 3 ft
1966	Imperial	3.6	Right-lateral slip, 1/2 in
1966	San Andreas	5.5	Known as the "Parkfield earthquake." Right-lateral slip of several inches. Maximum ground acceleration of 0.50 g—highest recorded to date.
1968	Coyote Creek	6.4	Right-lateral slip up to 15 in
1971	San Fernando	6.4–6.6	Known as the "San Fernando earthquake," or "Sylmar earthquake." Left-lateral slip up to 5 ft; north-side thrusting up 3 ft. Massive instrumentation due to 1965 Los Angeles building code resulted in more than 300 accelerograph plots. 1.24 g experienced at Pacoima dam.
1979	Imperial	6.6	Known as the "Imperial Valley earthquake." Right-lateral slip up to 55 cm with more than 30 km of surface rupture. Extensive accelerograph data collected. Resulted in the first accelerograph from an extensively damaged building (Imperial County Services Building).
1987	Whittier	6.1	Known as the "Whittier earthquake." Epicenter 10 mi east of downtown Los Angeles. 0.45 g maximum lateral acceleration; 0.20 g vertical acceleration typical. Strong shaking duration of 4 sec. Six fatalities; unreinforced masonry structures damaged significantly.
1989	Loma Prieta	7.1	Primarily noted as causing the collapse of the Oakland Interstate 880 Cypress structure and homes in the San Francisco Marina district, both due to soil amplification effects despite the large distance from the epicenter. Ground breakage at epicenter located in Santa Cruz Mountains; 62 fatalities.

APPENDIX C (continued)
Chronology of Important California Earthquakes

Date	Fault	Richter Magnitude	Surface Effects and Significance
1990	Upland	5.5	Acceleration of 0.23 g horizontally and 0.13 g vertically. Dozens of aftershocks. Occurred on the San Antonio Canyon fault, east of downtown Los Angeles. Most damage in Pomona; some damage to reinforced masonry structures.
1992	Yucca Valley	7.4	One fatality, hundreds of injuries. Buckled and displaced roads, damaged 1400 structures. 43-mi ground rupture. Followed by magnitude 6.5 quake at Big Bear resort area.
1992	Cape Mendocino	7.0	Offshore quake notable for largest-yet recorded ground acceleration of 1.85 g.
1994	Northridge	6.6	Previously unknown thrust fault. 40 s of shaking. Extensive damage to sections of the Santa Monica freeway and freeway overpasses not yet retrofitted. Unreinforced masonry buildings damaged, as expected. 60 fatalities, thousands of injuries.

PROFESSIONAL PUBLICATIONS, INC. ● Belmont, CA

APPENDIX D
Rigidity of Fixed Piers

(Calculated with $F = 100,000$, $t = 1.0$, and $E = 1,000,000$)

(h/d)	.00	.01	.02	.03	.04	.05	.06	.07	.08	.09
0.10	33.223	30.181	27.645	25.497	23.655	22.057	20.657	19.421	18.321	17.335
0.20	16.447	15.643	14.911	14.242	13.627	13.061	12.538	12.053	11.602	11.181
0.30	10.788	10.419	10.073	9.747	9.440	9.150	8.876	8.616	8.369	8.135
0.40	7.911	7.699	7.496	7.302	7.117	6.939	6.769	6.606	6.449	6.299
0.50	6.154	6.015	5.880	5.751	5.626	5.506	5.389	5.277	5.168	5.062
0.60	4.960	4.862	4.766	4.673	4.583	4.495	4.410	4.328	4.247	4.169
0.70	4.093	4.019	3.948	3.877	3.809	3.743	3.678	3.615	3.553	3.493
0.80	3.434	3.377	3.321	3.266	3.213	3.160	3.109	3.060	3.011	2.963
0.90	2.916	2.871	2.826	2.782	2.739	2.697	2.656	2.616	2.577	2.538
1.00	2.500	2.463	2.427	2.391	2.356	2.322	2.288	2.255	2.222	2.191
1.10	2.159	2.129	2.099	2.069	2.040	2.012	1.984	1.956	1.929	1.903
1.20	1.877	1.851	1.826	1.802	1.777	1.753	1.730	1.707	1.684	1.662
1.30	1.640	1.619	1.598	1.577	1.556	1.536	1.516	1.497	1.478	1.459
1.40	1.440	1.422	1.404	1.386	1.369	1.352	1.335	1.318	1.302	1.286
1.50	1.270	1.254	1.239	1.224	1.209	1.194	1.180	1.166	1.152	1.138
1.60	1.124	1.111	1.098	1.085	1.072	1.059	1.047	1.034	1.022	1.010
1.70	.999	.987	.976	.965	.954	.943	.932	.921	.911	.901
1.80	.890	.880	.870	.861	.851	.842	.832	.823	.814	.805
1.90	.796	.788	.779	.771	.762	.754	.746	.738	.730	.722
2.00	.714	.707	.699	.692	.685	.677	.670	.663	.656	.649
2.10	.643	.636	.629	.623	.617	.610	.604	.598	.592	.586
2.20	.580	.574	.568	.562	.557	.551	.546	.540	.535	.530
2.30	.525	.519	.514	.509	.504	.499	.495	.490	.485	.480
2.40	.476	.471	.467	.462	.458	.453	.449	.445	.441	.437
2.50	.432	.428	.424	.420	.417	.413	.409	.405	.401	.398
2.60	.394	.391	.387	.383	.380	.377	.373	.370	.367	.363
2.70	.360	.357	.354	.350	.347	.344	.341	.338	.335	.332
2.80	.330	.327	.324	.321	.318	.316	.313	.310	.307	.305
2.90	.302	.300	.297	.295	.292	.290	.287	.285	.283	.280
3.00	.278	.276	.273	.271	.269	.267	.264	.262	.260	.258
3.10	.256	.254	.252	.250	.248	.246	.244	.242	.240	.238
3.20	.236	.234	.232	.231	.229	.227	.225	.223	.222	.220
3.30	.218	.217	.215	.213	.212	.210	.208	.207	.205	.204
3.40	.202	.201	.199	.198	.196	.195	.193	.192	.190	.189
3.50	.187	.186	.185	.183	.182	.181	.179	.178	.177	.175
3.60	.174	.173	.172	.170	.169	.168	.167	.166	.164	.163
3.70	.162	.161	.160	.159	.157	.156	.155	.154	.153	.152
3.80	.151	.150	.149	.148	.147	.146	.145	.144	.143	.142
3.90	.141	.140	.139	.138	.137	.136	.135	.134	.133	.132
4.00	.132	.131	.130	.129	.128	.127	.126	.126	.125	.124
4.10	.123	.122	.122	.121	.120	.119	.118	.118	.117	.116
4.20	.115	.115	.114	.113	.112	.112	.111	.110	.110	.109
4.30	.108	.108	.107	.106	.106	.105	.104	.104	.103	.102
4.40	.102	.101	.100	.100	.099	.099	.098	.097	.097	.096
4.50	.096	.095	.094	.094	.093	.093	.092	.092	.091	.091
4.60	.090	.089	.089	.088	.088	.087	.087	.086	.086	.085
4.70	.085	.084	.084	.083	.083	.082	.082	.081	.081	.080
4.80	.080	.080	.079	.079	.078	.078	.077	.077	.076	.076
4.90	.076	.075	.075	.074	.074	.073	.073	.073	.072	.072
5.00	.071	.071	.071	.070	.070	.069	.069	.069	.068	.068
5.10	.068	.067	.067	.066	.066	.066	.065	.065	.065	.064
5.20	.064	.064	.063	.063	.063	.062	.062	.062	.061	.061
5.30	.061	.060	.060	.060	.059	.059	.059	.058	.058	.058
5.40	.058	.057	.057	.057	.056	.056	.056	.056	.055	.055
5.50	.055	.054	.054	.054	.054	.053	.053	.053	.052	.052
5.60	.052	.052	.051	.051	.051	.050	.050	.050	.050	.050
5.70	.049	.049	.049	.049	.048	.048	.048	.048	.048	.047
5.80	.047	.047	.047	.046	.046	.046	.046	.045	.045	.045
5.90	.045	.045	.044	.044	.044	.044	.044	.043	.043	.043
6.00	.043	.043	.042	.042	.042	.042	.042	.041	.041	.041
6.10	.041	.041	.040	.040	.040	.040	.040	.039	.039	.039
6.20	.039	.039	.039	.038	.038	.038	.038	.038	.038	.037
6.30	.037	.037	.037	.037	.037	.036	.036	.036	.036	.036
6.40	.036	.035	.035	.035	.035	.035	.035	.034	.034	.034
6.50	.034	.034	.034	.034	.033	.033	.033	.033	.033	.033
6.60	.033	.032	.032	.032	.032	.032	.032	.032	.031	.031
6.70	.031	.031	.031	.031	.031	.031	.030	.030	.030	.030
6.80	.030	.030	.030	.029	.029	.029	.029	.029	.029	.029
6.90	.029	.029	.028	.028	.028	.028	.028	.028	.028	.028
7.00	.027	.027	.027	.027	.027	.027	.027	.027	.027	.026
7.10	.026	.026	.026	.026	.026	.026	.026	.026	.026	.025
7.20	.025	.025	.025	.025	.025	.025	.025	.025	.025	.024
7.30	.024	.024	.024	.024	.024	.024	.024	.024	.024	.023
7.40	.023	.023	.023	.023	.023	.023	.023	.023	.023	.023
7.50	.023	.022	.022	.022	.022	.022	.022	.022	.022	.022
7.60	.022	.022	.021	.021	.021	.021	.021	.021	.021	.021
7.70	.021	.021	.021	.021	.021	.020	.020	.020	.020	.020
7.80	.020	.020	.020	.020	.020	.020	.020	.020	.019	.019
7.90	.019	.019	.019	.019	.019	.019	.019	.019	.019	.019
8.00	.019	.019	.019	.018	.018	.018	.018	.018	.018	.018
8.10	.018	.018	.018	.018	.018	.018	.018	.018	.017	.017
8.20	.017	.017	.017	.017	.017	.017	.017	.017	.017	.017

APPENDIX E
Rigidity of Cantilever Piers

(Calculated with $F = 100,000$, $t = 1.0$, and $E = 1,000,000$)

(h/d	.00	.01	.02	.03	.04	.05	.06	.07	.08	.09
0.10	32.895	29.822	27.255	25.076	23.203	21.575	20.146	18.880	17.752	16.738
0.20	15.823	14.992	14.233	13.538	12.898	12.308	11.761	11.252	10.778	10.335
0.30	9.921	9.531	9.165	8.820	8.495	8.187	7.895	7.618	7.356	7.106
0.40	6.868	6.642	6.425	6.219	6.021	5.833	5.652	5.479	5.313	5.153
0.50	5.000	4.853	4.712	4.576	4.445	4.319	4.197	4.080	3.968	3.859
0.60	3.754	3.652	3.555	3.460	3.369	3.280	3.195	3.112	3.032	2.955
0.70	2.880	2.808	2.738	2.670	2.604	2.540	2.478	2.418	2.360	2.303
0.80	2.248	2.195	2.143	2.093	2.045	1.997	1.952	1.907	1.864	1.822
0.90	1.781	1.741	1.702	1.665	1.628	1.593	1.558	1.524	1.492	1.460
1.00	1.429	1.398	1.369	1.340	1.312	1.285	1.259	1.233	1.208	1.183
1.10	1.160	1.136	1.114	1.092	1.070	1.049	1.028	1.008	.989	.970
1.20	.951	.933	.916	.898	.881	.865	.849	.833	.818	.803
1.30	.788	.774	.760	.746	.733	.720	.707	.695	.683	.671
1.40	.659	.648	.636	.626	.615	.604	.594	.584	.575	.565
1.50	.556	.546	.537	.529	.520	.512	.503	.495	.487	.480
1.60	.472	.465	.457	.450	.443	.436	.430	.423	.417	.410
1.70	.404	.398	.392	.386	.380	.375	.369	.364	.358	.353
1.80	.348	.343	.338	.333	.329	.324	.319	.315	.310	.306
1.90	.302	.298	.294	.290	.286	.282	.278	.274	.270	.267
2.00	.263	.260	.256	.253	.250	.246	.243	.240	.237	.234
2.10	.231	.228	.225	.222	.219	.216	.214	.211	.208	.206
2.20	.203	.201	.198	.196	.194	.191	.189	.187	.184	.182
2.30	.180	.178	.176	.174	.172	.170	.168	.166	.164	.162
2.40	.160	.158	.156	.155	.153	.151	.149	.148	.146	.145
2.50	.143	.141	.140	.138	.137	.135	.134	.132	.131	.129
2.60	.128	.127	.125	.124	.123	.121	.120	.119	.118	.116
2.70	.115	.114	.113	.112	.111	.109	.108	.107	.106	.105
2.80	.104	.103	.102	.101	.100	.099	.098	.097	.096	.095
2.90	.094	.093	.092	.091	.091	.090	.089	.088	.087	.086
3.00	.086	.085	.084	.083	.082	.082	.081	.080	.079	.079
3.10	.078	.077	.076	.076	.075	.074	.074	.073	.072	.072
3.20	.071	.071	.070	.069	.069	.068	.067	.067	.066	.066
3.30	.065	.065	.064	.063	.063	.062	.062	.061	.061	.060
3.40	.060	.059	.059	.058	.058	.057	.057	.056	.056	.055
3.50	.055	.055	.054	.054	.053	.053	.052	.052	.052	.051
3.60	.051	.050	.050	.050	.049	.049	.048	.048	.048	.047
3.70	.047	.046	.046	.046	.045	.045	.045	.044	.044	.044
3.80	.043	.043	.043	.042	.042	.042	.041	.041	.041	.040
3.90	.040	.040	.040	.039	.039	.039	.038	.038	.038	.038
4.00	.037	.037	.037	.037	.036	.036	.036	.035	.035	.035
4.10	.035	.034	.034	.034	.034	.033	.033	.033	.033	.033
4.20	.032	.032	.032	.032	.031	.031	.031	.031	.031	.030
4.30	.030	.030	.030	.030	.029	.029	.029	.029	.029	.028
4.40	.028	.028	.028	.028	.028	.027	.027	.027	.027	.027
4.50	.026	.026	.026	.026	.026	.026	.025	.025	.025	.025
4.60	.025	.025	.024	.024	.024	.024	.024	.024	.024	.023
4.70	.023	.023	.023	.023	.023	.023	.023	.022	.022	.022
4.80	.022	.022	.022	.021	.021	.021	.021	.021	.021	.021
4.90	.021	.020	.020	.020	.020	.020	.020	.020	.020	.020
5.00	.019	.019	.019	.019	.019	.019	.019	.019	.019	.018
5.10	.018	.018	.018	.018	.018	.018	.018	.018	.017	.017
5.20	.017	.017	.017	.017	.017	.017	.017	.017	.017	.016
5.30	.016	.016	.016	.016	.016	.016	.016	.016	.016	.016
5.40	.015	.015	.015	.015	.015	.015	.015	.015	.015	.015
5.50	.015	.015	.015	.014	.014	.014	.014	.014	.014	.014
5.60	.014	.014	.014	.014	.014	.014	.013	.013	.013	.013
5.70	.013	.013	.013	.013	.013	.013	.013	.013	.013	.013
5.80	.013	.012	.012	.012	.012	.012	.012	.012	.012	.012
5.90	.012	.012	.012	.012	.012	.012	.012	.012	.011	.011
6.00	.011	.011	.011	.011	.011	.011	.011	.011	.011	.011
6.10	.011	.011	.011	.011	.011	.011	.010	.010	.010	.010
6.20	.010	.010	.010	.010	.010	.010	.010	.010	.010	.010
6.30	.010	.010	.010	.010	.010	.010	.010	.009	.009	.009
6.40	.009	.009	.009	.009	.009	.009	.009	.009	.009	.009
6.50	.009	.009	.009	.009	.009	.009	.009	.009	.009	.009
6.60	.009	.009	.008	.008	.008	.008	.008	.008	.008	.008
6.70	.008	.008	.008	.008	.008	.008	.008	.008	.008	.008
6.80	.008	.008	.008	.008	.008	.008	.008	.008	.008	.008
6.90	.007	.007	.007	.007	.007	.007	.007	.007	.007	.007
7.00	.007	.007	.007	.007	.007	.007	.007	.007	.007	.007
7.10	.007	.007	.007	.007	.007	.007	.007	.007	.007	.007
7.20	.007	.007	.007	.007	.006	.006	.006	.006	.006	.006
7.30	.006	.006	.006	.006	.006	.006	.006	.006	.006	.006
7.40	.006	.006	.006	.006	.006	.006	.006	.006	.006	.006
7.50	.006	.006	.006	.006	.006	.006	.006	.006	.006	.006
7.60	.006	.006	.006	.006	.006	.006	.005	.005	.005	.005
7.70	.005	.005	.005	.005	.005	.005	.005	.005	.005	.005
7.80	.005	.005	.005	.005	.005	.005	.005	.005	.005	.005
7.90	.005	.005	.005	.005	.005	.005	.005	.005	.005	.005
8.00	.005	.005	.005	.005	.005	.005	.005	.005	.005	.005
8.10	.005	.005	.005	.005	.005	.005	.005	.005	.005	.005
8.20	.004	.004	.004	.004	.004	.004	.004	.004	.004	.004

PROFESSIONAL PUBLICATIONS, INC. ● Belmont, CA

APPENDIX F
Accelerogram of 1940 El Centro Earthquake

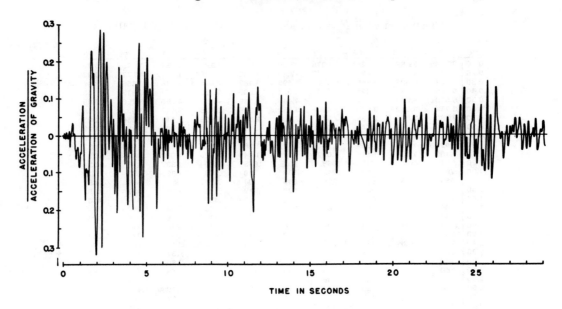

Reproduced from Donald E. Hudson, "Ground Motion Measurements," in *Earthquake Engineering*, Robert L. Wiegel, ed., © 1970, p. 113. Reprinted by permission of Prentice-Hall, Inc., Englewood Cliffs, NJ.

APPENDIX G
Bibliography for Further Reading

Ambrose, James, and Dimitry Vergun, *Simplified Building Design for Wind and Earthquake Forces*, New York: John Wiley, 1980.

Amrhein, J.E., *Reinforced Masonry Engineering Handbook*, 2d ed., Los Angeles: Masonry Institute of America, 1973.

Architectural Institute of Japan, ed., *Design Essentials in Earthquake Resistant Buildings*, New York: Elsevier, 1970.

Blume, John A., Nathan M. Newmark, and Leo H. Corning, *Design of Multistory Reinforced Concrete Buildings for Earthquake Motions*, Skokie, Ill.: Portland Cement Association, 1961.

Breyer, Donald E., *Design of Wood Structures*, 2d ed., New York: McGraw-Hill, 1980.

Building Code Requirements for Reinforced Concrete (ACI-318-89), Detroit: American Concrete Institute, 1989.

Department of the Army, et al., *Seismic Design Guidelines for Essential Buildings* (TM 5-809-10-1), Washington, D.C., 1986.

Dowrick, D.J., *Earthquake Resistant Design*, 2d ed., New York: John Wiley, 1987.

Federal Highway Administration, *Seismic Design of Highway Bridges: Workshop Manual* (FHWA-IP-81-2), Washington, D.C., 1981.

Gould, Phillip L., and Salman H. Abu-Sitta, *Dynamic Response of Structures To Wind and Earthquake Loading*, New York: John Wiley, 1980.

Green, Norman B., *Earthquake Resistant Building Design and Construction*, New York: Reinhold, 1981.

Hamilton, Warren, "Plate Tectonics and Man," *USGS Annual Report, Fiscal Year 1966* (reprint no. 1978-261-227/27). Washington, D.C.: U.S. Government Printing Office, 1978.

Iacopi, Robert, *Earthquake Country*, Menlo Park, Calif.: Lane, 1964.

Merritt, Frederic S., *Structural Steel Designers' Handbook*, New York: McGraw-Hill, 1972.

Murty, T. S., *Tsunamis: Seismic Sea Waves*, Ottawa, Canada: Department of Fisheries and the Environment, 1977.

Naeim, Farzad, ed., *The Seismic Design Handbook*, New York: Reinhold, 1989.

Popov, Egor P., "Seismic Steel Framing Systems for Tall Buildings," *Engineering Journal of the American Institute of Steel Construction*, 19, no. 3, 1982.

Recommended Lateral Force Requirements and Commentary, San Francisco: Structural Engineers Association of California, 1990.

Rosenblueth, Emilio, ed., *Design of Earthquake Resistant Structures*, New York: John Wiley, 1980.

Seto, William W., *Mechanical Vibrations*, New York: McGraw-Hill, 1964.

Stratta, James L., *Manual of Seismic Design*, Englewood Cliffs, N.J.: Prentice-Hall, 1987.

Teal, Edward J., "Seismic Design Practice for Steel Buildings," *Engineering Journal* (American Institute of Steel Construction), 12, no. 4 (1975).

Uniform Building Code, Whittier, CA: International Conference of Building Officials, 1991.

United States Geological Survey, "Active Faults in California" (1977-0-240-96 6/48). Washington, D.C.: U.S. Government Printing Office, 1977.

—. "Earthquakes" (1979-311-348/11). Washington, D.C.: U.S. Government Printing Office, 1979.

—. "The Severity of an Earthquake," (024-001-03204-0) (1979-0-281-363[33]). Washington, D.C.: U.S. Government Printing Office, 1979.

Wakabayashi, Minoru, *Design of Earthquake-Resistant Buildings*, New York: McGraw-Hill, 1986.

Wiegel, Robert L., ed., *Earthquake Engineering*, Englewood Cliffs, N.J.: Prentice-Hall, 1970.

Yanev, Peter, *Peace of Mind in Earthquake Country*, Reprint. San Francisco: Chronicle Books, 1980.

APPENDIX H
Summary of California Special Seismic Exam Question Format

At its June 15, 1990 meeting, the California State Board of Registration for Professional Engineers and Land Surveyors adopted the following CTB/McGraw-Hill proposal for exam format for its special seismic exam. These tasks were derived from an analysis of approximately 800 responses to a CTB/McGraw-Hill task analysis questionnaire. The length of time allowed to complete the seismic and land surveying problems was increased by the Board from two hours to four hours.

An entry-level civil engineer must demonstrate the knowledge, skills, and abilities necessary to carry out the following responsibilities as they relate to seismic principles.

Subject	Description	Percentage of Test Points
1	Understand the purpose and scope of the seismic aspects of a project.	11.11%
2	Compare and evaluate lateral load resistant systems and identify appropriate systems.	7.80%
3	Identify applicable codes and regulatory requirements relative to the specific site conditions and the structure.	10.16%
4	Assess the need for and the application of more stringent standards than the code and regulatory requirements.	4.92%
5	Understand how fields or disciplines, such as engineering, geology, architecture, and construction, interrelate with seismic design considerations.	9.19%
6	Consider time and cost factors in the design and construction of the lateral load resistant systems.	5.48%
7	Identify and evaluate existing sources of seismic-related information, such as soils reports, site seismicity, and computer programs, and ascertain whether additional information is necessary.	8.56%
8	Understand seismic principles as applied to civil engineering analysis, design, and construction.	13.21%
9	Perform the necessary analysis, engineering computations, and development details to solve civil engineering design problems relative to seismic effects on structures and other fixed works.	9.84%
10	Monitor the construction of seismic design aspects of the project for quality assurance and quality control.	7.53%
11	Recognize the extent and limits of the civil engineering designer's area of competency and authority, and the interrelationship with other civil engineering areas of competency and authority.	12.21%

PROFESSIONAL PUBLICATIONS, INC. ● Belmont, CA

APPENDIX I
Subjects on the California Special Seismic Examination

In 1988, the California Seismic Safety Commission recommended to the California Board of Registration for Professional Engineers that the special seismic examination cover the subjects listed below.[a] While it is seemingly impossible to predict what subjects will be covered in any particular examination, over the long run, question topics have generally coincided with the recommendations.

1. Recognize when existing buildings are nonconforming, and hence, may require retrofitting to ensure safety.

   - buildings with unreinforced masonry bearing walls
   - buildings with unreinforced masonry infill walls
   - buildings with nonductile frames
   - tilt-up buildings
   - buildings with soft stories and discontinuities
   - buildings with parapets and appendages
   - deteriorated buildings with cumulative damage, including excessive settlement

2. Be familiar with basic seismological knowledge.

   - magnitude and intensity scales
   - response spectra
   - attenuation (versus distance from fault) relationships
   - geotechnical information needed to evaluate sites
   - effect of different soil types on building response
   - out-of-phase motion of tall structures
   - effects of ground failure (liquefaction, lateral spreading, and fault rupture)

3. Know issues specific to California.

   - types of faults
   - earthquake history
   - seismic zones
   - California law
     - Field Act
     - Hospital Seismic Safety Act
     - Essential Services Facilities Act
     - Title 24

4. Have basic knowledge of structural issues.

   - use of and need for different structural systems (e.g., bearing walls, semi-ductile frames)
   - ductility concepts
   - beam-column connections
   - transfer of lateral force to supporting members
   - effects of building form and size on response
   - effects of structural irregularity

5. Be able to use the UBC.

   - philosophy behind the UBC provisions (i.e., life safety)
   - UBC response spectra
   - determination of proper R_w values
   - determination of lateral force
   - drift limitations
   - limitations of the UBC (consequences of relying only on the UBC and not recognizing local site problems such as fault rupture and sliding)
   - understanding of why buildings with different structural systems are designed to different lateral forces (i.e., R_w values versus energy absorption method)

---

[a]Letter dated October 20, 1988 from L. Thomas Tobin, Executive Director of the SSC to Ms. Darlene Stroup, Executive Office of BORPELS. (Subjects listed have been edited and reorganized.)

APPENDIX J
Standard Welding Symbols of the AISC/AWS

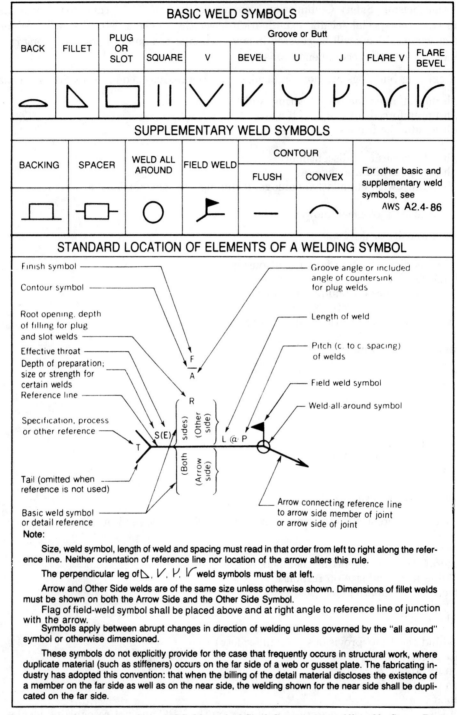

Note:

Size, weld symbol, length of weld and spacing must read in that order from left to right along the reference line. Neither orientation of reference line nor location of the arrow alters this rule.

The perpendicular leg of ⊿, V, V, V weld symbols must be at left.

Arrow and Other Side welds are of the same size unless otherwise shown. Dimensions of fillet welds must be shown on both the Arrow Side and the Other Side Symbol.

Flag of field-weld symbol shall be placed above and at right angle to reference line of junction with the arrow.

Symbols apply between abrupt changes in direction of welding unless governed by the "all around" symbol or otherwise dimensioned.

These symbols do not explicitly provide for the case that frequently occurs in structural work, where duplicate material (such as stiffeners) occurs on the far side of a web or gusset plate. The fabricating industry has adopted this convention: that when the billing of the detail material discloses the existence of a member on the far side as well as on the near side, the welding shown for the near side shall be duplicated on the far side.

Reprinted with permission from AISC *Manual of Steel Construction, Allowable Stress Design,*
9th ed., © 1989, American Institute of Steel Construction.

APPENDIX K
Commercial Wood Framing Straps, Anchors, and Holdowns
PA/HPA Series Purlin Anchors

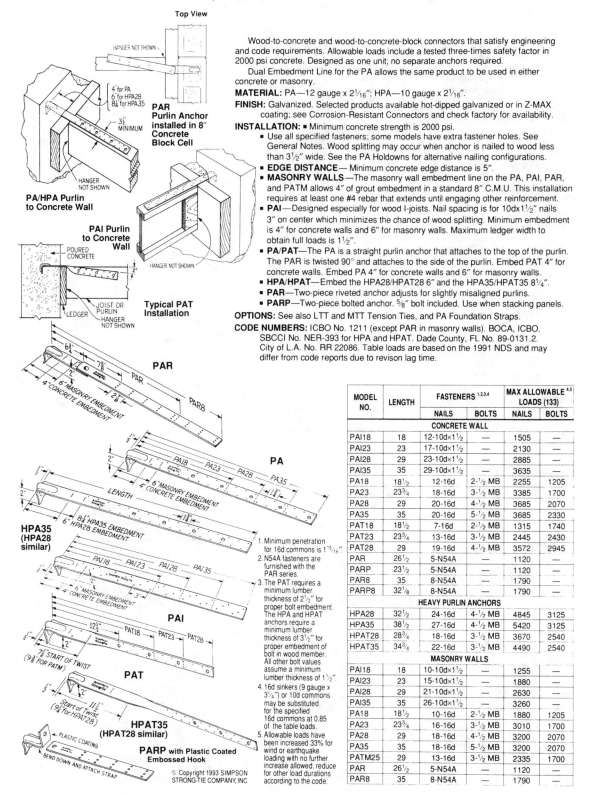

Wood-to-concrete and wood-to-concrete-block connectors that satisfy engineering and code requirements. Allowable loads include a tested three-times safety factor in 2000 psi concrete. Designed as one unit; no separate anchors required.

Dual Embedment Line for the PA allows the same product to be used in either concrete or masonry.

MATERIAL: PA—12 gauge x $2\frac{1}{16}$"; HPA—10 gauge x $2\frac{1}{16}$".

FINISH: Galvanized. Selected products available hot-dipped galvanized or in Z-MAX coating; see Corrosion-Resistant Connectors and check factory for availability.

INSTALLATION: ■ Minimum concrete strength is 2000 psi.
- Use all specified fasteners; some models have extra fastener holes. See General Notes. Wood splitting may occur when anchor is nailed to wood less than $3\frac{1}{2}$" wide. See the PA Holdowns for alternative nailing configurations.
- **EDGE DISTANCE**— Minimum concrete edge distance is 5".
- **MASONRY WALLS**—The masonry wall embedment line on the PA, PAI, PAR, and PATM allows 4" of grout embedment in a standard 8" C.M.U. This installation requires at least one #4 rebar that extends until engaging other reinforcement.
- **PAI**—Designed especially for wood I-joists. Nail spacing is for 10dx$1\frac{1}{2}$" nails 3" on center which minimizes the chance of wood splitting. Minimum embedment is 4" for concrete walls and 6" for masonry walls. Maximum ledger width to obtain full loads is $1\frac{1}{2}$".
- **PA/PAT**—The PA is a straight purlin anchor that attaches to the top of the purlin. The PAR is twisted 90° and attaches to the side of the purlin. Embed PAT 4" for concrete walls. Embed PA 4" for concrete walls and 6" for masonry walls.
- **HPA/HPAT**—Embed the HPA28/HPAT28 6" and the HPA35/HPAT35 $8\frac{1}{4}$".
- **PAR**—Two-piece riveted anchor adjusts for slightly misaligned purlins.
- **PARP**—Two-piece bolted anchor. $\frac{5}{8}$" bolt included. Use when stacking panels.

OPTIONS: See also LTT and MTT Tension Ties, and PA Foundation Straps.

CODE NUMBERS: ICBO No. 1211 (except PAR in masonry walls). BOCA, ICBO, SBCCI No. NER-393 for HPA and HPAT. Dade County, FL No. 89-0131.2. City of L.A. No. RR 22086. Table loads are based on the 1991 NDS and may differ from code reports due to revison lag time.

1. Minimum penetration for 16d commons is $1\frac{15}{16}$".
2. N54A fasteners are furnished with the PAR series.
3. The PAT requires a minimum lumber thickness of $2\frac{1}{2}$" for proper bolt embedment. The HPA and HPAT anchors require a minimum lumber thickness of $3\frac{1}{2}$" for proper embedment of bolt in wood member. All other bolt values assume a minimum lumber thickness of $1\frac{1}{2}$".
4. 16d sinkers (9 gauge x $3\frac{1}{4}$") or 10d commons may be substituted for the specified 16d commons at 0.85 of the table loads.
5. Allowable loads have been increased 33% for wind or earthquake loading with no further increase allowed; reduce for other load durations according to the code.

MODEL NO.	LENGTH	FASTENERS [1,2,3,4]		MAX ALLOWABLE [4,5] LOADS (133)	
		NAILS	BOLTS	NAILS	BOLTS
CONCRETE WALL					
PAI18	18	12-10dx$1\frac{1}{2}$	—	1505	—
PAI23	23	17-10dx$1\frac{1}{2}$	—	2130	—
PAI28	29	23-10dx$1\frac{1}{2}$	—	2885	—
PAI35	35	29-10dx$1\frac{1}{2}$	—	3635	—
PA18	$18\frac{1}{2}$	12-16d	2-$\frac{1}{2}$ MB	2255	1205
PA23	$23\frac{3}{4}$	18-16d	3-$\frac{1}{2}$ MB	3385	1700
PA28	29	20-16d	4-$\frac{1}{2}$ MB	3685	2070
PA35	35	20-16d	5-$\frac{1}{2}$ MB	3685	2330
PAT18	$18\frac{1}{2}$	7-16d	2-$\frac{1}{2}$ MB	1315	1740
PAT23	$23\frac{3}{4}$	13-16d	3-$\frac{1}{2}$ MB	2445	2430
PAT28	29	19-16d	4-$\frac{1}{2}$ MB	3572	2945
PAR	$26\frac{1}{2}$	5-N54A	—	1120	—
PARP	$23\frac{1}{2}$	5-N54A	—	1120	—
PAR8	35	8-N54A	—	1790	—
PARP8	$32\frac{1}{8}$	8-N54A	—	1790	—
HEAVY PURLIN ANCHORS					
HPA28	$32\frac{1}{2}$	24-16d	4-$\frac{1}{2}$ MB	4845	3125
HPA35	$38\frac{1}{2}$	27-16d	4-$\frac{1}{2}$ MB	5420	3125
HPAT28	$28\frac{3}{4}$	18-16d	3-$\frac{1}{2}$ MB	3670	2540
HPAT35	$34\frac{3}{4}$	22-16d	3-$\frac{1}{2}$ MB	4490	2540
MASONRY WALLS					
PAI18	18	10-10dx$1\frac{1}{2}$	—	1255	—
PAI23	23	15-10dx$1\frac{1}{2}$	—	1880	—
PAI28	29	21-10dx$1\frac{1}{2}$	—	2630	—
PAI35	35	26-10dx$1\frac{1}{2}$	—	3260	—
PA18	$18\frac{1}{2}$	10-16d	2-$\frac{1}{2}$ MB	1880	1205
PA23	$23\frac{3}{4}$	16-16d	3-$\frac{1}{2}$ MB	3010	1700
PA28	29	18-16d	4-$\frac{1}{2}$ MB	3200	2070
PA35	35	18-16d	5-$\frac{1}{2}$ MB	3200	2070
PATM25	29	13-16d	3-$\frac{1}{2}$ MB	2335	1700
PAR	$26\frac{1}{2}$	5-N54A	—	1120	—
PAR8	35	8-N54A	—	1790	—

© Copyright 1993 SIMPSON STRONG-TIE COMPANY, INC.

Reprinted with permission from *Connectors for Wood Construction*,
Product & Information Manual © 1993, Simpson Strong-Tie Company, Inc.

APPENDIX K (continued)
HDA/HD Holdowns

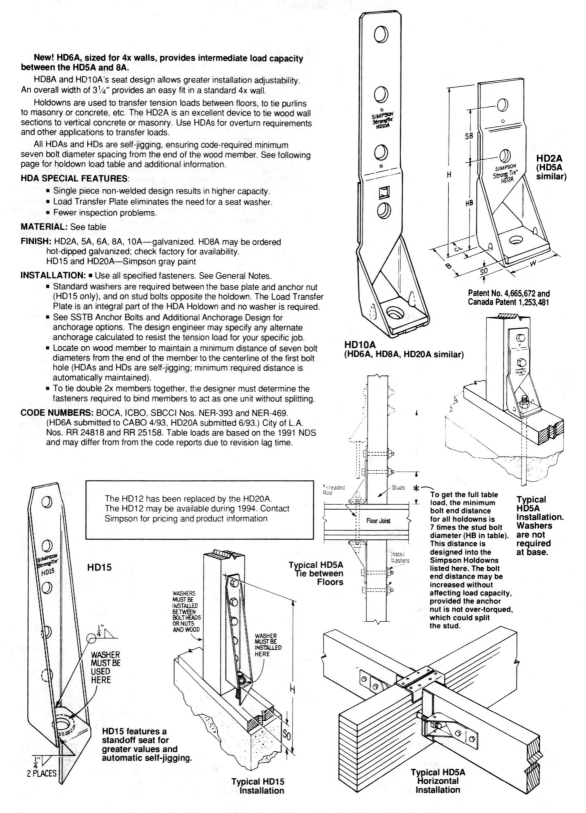

New! HD6A, sized for 4x walls, provides intermediate load capacity between the HD5A and 8A.

HD8A and HD10A's seat design allows greater installation adjustability. An overall width of 3¼" provides an easy fit in a standard 4x wall.

Holdowns are used to transfer tension loads between floors, to tie purlins to masonry or concrete, etc. The HD2A is an excellent device to tie wood wall sections to vertical concrete or masonry. Use HDAs for overturn requirements and other applications to transfer loads.

All HDAs and HDs are self-jigging, ensuring code-required minimum seven bolt diameter spacing from the end of the wood member. See following page for holdown load table and additional information.

HDA SPECIAL FEATURES:
- Single piece non-welded design results in higher capacity.
- Load Transfer Plate eliminates the need for a seat washer.
- Fewer inspection problems.

MATERIAL: See table

FINISH: HD2A, 5A, 6A, 8A, 10A—galvanized. HD8A may be ordered hot-dipped galvanized; check factory for availability. HD15 and HD20A—Simpson gray paint

INSTALLATION: ■ Use all specified fasteners. See General Notes.
- Standard washers are required between the base plate and anchor nut (HD15 only), and on stud bolts opposite the holdown. The Load Transfer Plate is an integral part of the HDA Holdown and no washer is required.
- See SSTB Anchor Bolts and Additional Anchorage Design for anchorage options. The design engineer may specify any alternate anchorage calculated to resist the tension load for your specific job.
- Locate on wood member to maintain a minimum distance of seven bolt diameters from the end of the member to the centerline of the first bolt hole (HDAs and HDs are self-jigging; minimum required distance is automatically maintained).
- To tie double 2x members together, the designer must determine the fasteners required to bind members to act as one unit without splitting.

CODE NUMBERS: BOCA, ICBO, SBCCI Nos. NER-393 and NER-469. (HD6A submitted to CABO 4/93, HD20A submitted 6/93.) City of L.A. Nos. RR 24818 and RR 25158. Table loads are based on the 1991 NDS and may differ from from the code reports due to revision lag time.

HD10A
(HD6A, HD8A, HD20A similar)

HD2A
(HD5A similar)

Patent No. 4,665,672 and Canada Patent 1,253,481

The HD12 has been replaced by the HD20A. The HD12 may be available during 1994. Contact Simpson for pricing and product information

HD15

WASHER MUST BE USED HERE

2 PLACES

HD15 features a standoff seat for greater values and automatic self-jigging.

Typical HD15 Installation

WASHERS MUST BE INSTALLED BETWEEN BOLT HEADS OR NUTS AND WOOD

WASHER MUST BE INSTALLED HERE

Typical HD5A Tie between Floors

Threaded Rod

Floor Joist

Studs

Install Washers

To get the full table load, the minimum bolt end distance for all holdowns is 7 times the stud bolt diameter (HB in table). This distance is designed into the Simpson Holdowns listed here. The bolt end distance may be increased without affecting load capacity, provided the anchor nut is not over-torqued, which could split the stud.

Typical HD5A Installation. Washers are not required at base.

Typical HD5A Horizontal Installation

Reprinted with permission from *Connectors for Wood Construction*, Product & Information Manual © 1993, Simpson Strong-Tie Company, Inc.

APPENDIX K (continued)
HDA/HD Holdowns

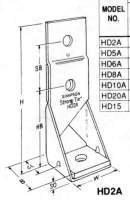

HD2A

MODEL NO.	MATERIAL		DIMENSIONS							FASTENERS		AVG ULT	ALLOWABLE LOADS[1,2] (133)					
	BASE	BODY	HB [4]	SB	W	H	B	SO	CL	ANCHOR DIA	STUD BOLTS		LENGTH OF BOLT [3,5,6] IN WOOD MEMBER					
													$1\frac{1}{2}$	2	$2\frac{1}{2}$	3	$3\frac{1}{2}$	$5\frac{1}{2}$
HD2A	7 ga	12 ga	$4\frac{1}{2}$	$2\frac{1}{2}$	$2\frac{1}{2}$	8	$2\frac{5}{8}$	$\frac{1}{4}$	$1\frac{1}{2}$	$\frac{5}{8}$	$2\text{-}\frac{5}{8}$	12150	1555	2055	2565	2775	2775	2760
HD5A	3 ga	10 ga	$5\frac{1}{4}$	3	$3\frac{3}{16}$	$9\frac{3}{8}$	$3\frac{1}{2}$	$\frac{1}{2}$	$2\frac{1}{16}$	$\frac{3}{4}$	$2\text{-}\frac{3}{4}$	20767	1870	2485	3095	3705	4010	3980
HD6A	$\frac{3}{8}$	7 ga	$6\frac{1}{8}$	$3\frac{1}{2}$	$3\frac{1}{4}$	$11\frac{1}{4}$	$3\frac{7}{16}$	$\frac{3}{4}$	$2\frac{1}{16}$	$\frac{7}{8}$	$2\text{-}\frac{7}{8}$	27333	2275	2985	3685	4405	5100	5510
HD8A	$\frac{3}{8}$	7 ga	$6\frac{1}{8}$	$3\frac{1}{2}$	$3\frac{1}{4}$	$14\frac{3}{4}$	$3\frac{7}{16}$	$\frac{3}{4}$	$2\frac{1}{16}$	$\frac{7}{8}$	$3\text{-}\frac{7}{8}$	28667	3220	4350	5415	6465	7460	7910
HD10A	$\frac{3}{8}$	7 ga	$6\frac{1}{8}$	$3\frac{1}{2}$	$3\frac{1}{4}$	18	$3\frac{7}{16}$	$\frac{3}{4}$	$2\frac{1}{16}$	$\frac{7}{8}$	$4\text{-}\frac{7}{8}$	28667	3945	5540	6935	8310	10425	9900
HD20A	$\frac{3}{8}$	3 ga	7	4	$3\frac{1}{2}$	$20\frac{3}{4}$	$3\frac{5}{8}$	$\frac{5}{8}$	$2\frac{3}{16}$	1	4-1	51233	—	—	—	—	11080	13380
HD15	$\frac{3}{8}$	3 ga	7	4	$3\frac{1}{2}$	$24\frac{1}{2}$	$4\frac{1}{4}$	$3\frac{5}{8}$	$2\frac{1}{8}$	$1\frac{1}{4}$	5-1	43750	—	—	—	—	—	15305

1. Allowable loads are based on the lower of (a) the bolt values in accordance with the NDS, and (b) the ultimate load on a steel test jig divided by 2.5.
2. Allowable loads have been increased 33% for wind or earthquake loading with no further increase allowed; reduce where other load durations govern.
3. The wood member must be sized for the load-carrying capacity at the critical net section, reducing the gross section area for holes or other removed wood as specified in the code.
4. HB is the required minimum distance from the end of the stud to the center of the first stud bolt hole. End distance may be increased as necessary for installation.
5. HD15 requires a minimum 6x6 nominal post.
6. Use a minimum 4x6 stud for the HD20A.
7. The anchor embedment and configuration must be specified. See SSTB Anchor Bolts and Additional Anchor Designs.

APPENDIX K (continued)
ST/FHA/PS Strap Ties

Over-notching problems? Piling Strap connects wood pilings to floor girders in elevated structures. Hot-dipped galvanized for greater corrosion resistance.

Install Strap Ties where plates or soles are cut, at wall intersections, and as ridge ties and truss plates. LSTA and MSTA straps are engineered for use on 1½″ members. The 3″ center-to-center nail spacing reduces the possibility of splitting. For the MST, this may be a problem on lumber narrower than 3½″; either fill every nail hole with 10dx1½″ nails or fill every other nail hole with 16d commons. Reduce the allowable load based on the size and quantity of fasteners used.

FINISH: HST— Simpson gray paint; PS and MST—Hot-dipped galvanized; all others—Galvanized. Selected products available in stainless steel or Z-MAX; see Corrosion-Resistant Connectors.

INSTALLATION: Use all specified fasteners. See General Notes.

OPTIONS: ▪ Special sizes can be made to order.
▪ See also HCST.

CODE NUMBERS: BOCA, ICBO, and SBCCI Nos. NER-413 and NER-443, except MST72 and MSTI72; Dade County, FL No. 89-0131.4. City of L.A. No. RR 25119. Table loads are based on the 1991 NDS and may differ from code reports due to revision lag time.

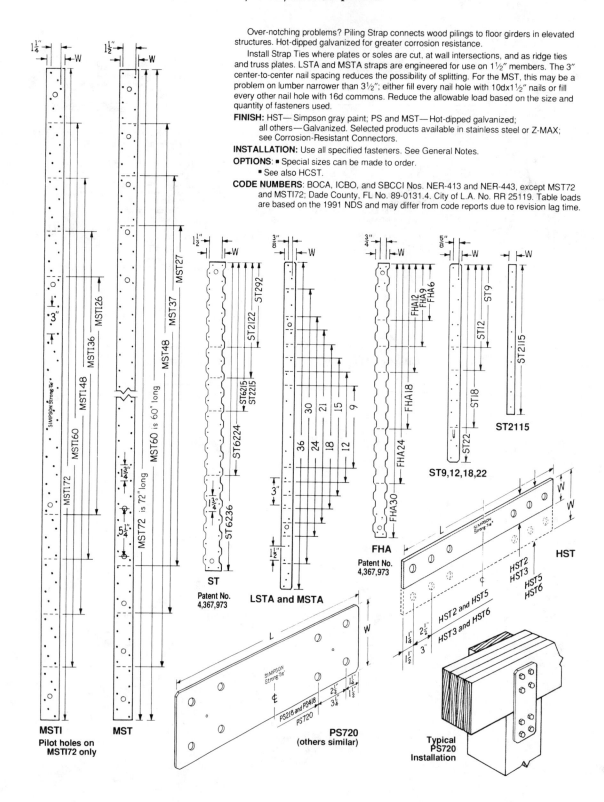

Reprinted with permission from *Connectors for Wood Construction*,
Product & Information Manual © 1993, Simpson Strong-Tie Company, Inc.

APPENDIX K (continued)
ST/FHA/PS Strap Ties

MODEL NO.	MATL	DIMENSIONS		FASTENERS (TOTAL)		ALLOWABLE LOADS			
						NAILS		BOLTS[3]	
		W	L	NAILS[2]	BOLTS	FLOOR (100)	MAX[1] (133)	FLOOR (100)	MAX[1] (133)
LSTA9	20 ga	1¼	9	8-10d	—	440	585	—	—
LSTA12	20 ga	1¼	12	10-10d	—	555	740	—	—
LSTA15	20 ga	1¼	15	12-10d	—	670	895	—	—
LSTA18	20 ga	1¼	18	14-10d	—	790	1055	—	—
LSTA21	20 ga	1¼	21	16-10d	—	905	1205	—	—
LSTA24	20 ga	1¼	24	18-10d	—	975	1295	—	—
LSTA30	18 ga	1¼	30	22-10d	—	1245	1640	—	—
LSTA36	18 ga	1¼	36	26-10d	—	1290	1640	—	—
MSTA9	18 ga	1¼	9	8-10d	—	430	570	—	—
MSTA12	18 ga	1¼	12	10-10d	—	545	725	—	—
MSTA15	18 ga	1¼	15	12-10d	—	660	880	—	—
MSTA18	18 ga	1¼	18	14-10d	—	775	1035	—	—
MSTA21	18 ga	1¼	21	16-10d	—	895	1195	—	—
MSTA24	18 ga	1¼	24	18-10d	—	1015	1355	—	—
MSTA30	16 ga	1¼	30	22-10d	—	1255	1670	—	—
MSTA36	16 ga	1¼	36	26-10d	—	1495	1995	—	—
ST292	20 ga	2 1/16	9 5/16	12-16d	—	790	1055	—	—
ST2122	20 ga	2 1/16	12 13/16	16-16d	—	1070	1425	—	—
ST2115	20 ga	¾	16 5/16	10-16d	—	450	600	—	—
ST2215	20 ga	2 1/16	16 5/16	20-16d	—	1210	1615	—	—
ST6215	16 ga	2 1/16	16 5/16	20-16d	—	1330	1775	—	—
ST6224	16 ga	2 1/16	23 5/16	28-16d	—	1890	2520	—	—
ST6236	14 ga	2 1/16	33 13/16	40-16d	—	2475	3300	—	—
ST9	16 ga	1¼	9	8-16d	—	540	720	—	—
ST12	16 ga	1¼	11⅝	10-16d	—	675	900	—	—
ST18	16 ga	1¼	17¾	14-16d	—	945	1260	—	—
ST22	16 ga	1¼	21⅝	18-16d	—	1215	1620	—	—
FHA6	12 ga	1 7/16	6	8-16d	—	550	735	—	—
FHA9	12 ga	1 7/16	9	8-16d	—	550	735	—	--
FHA12	12 ga	1 7/16	11⅝	8-16d	—	550	735	—	—
FHA18	12 ga	1 7/16	17¾	8-16d	—	550	735	—	—
FHA24	12 ga	1 7/16	23⅞	8-16d	—	550	735	—	—
FHA30	12 ga	1 7/16	30	8-16d	—	550	735	—	—
MSTI26	12 ga	2 1/16	26	26-10dx1½	—	1130	1510	—	—
MSTI36	12 ga	2 1/16	36	36-10dx1½	—	1565	2090	—	—
MSTI48	12 ga	2 1/16	48	48-10dx1½	—	2135	2850	—	—
MSTI60	12 ga	2 1/16	60	60-10dx1½	—	2760	3680	—	—
MSTI72	12 ga	2 1/16	72	72-10dx1½	—	3310	4415	—	—
MST27	12 ga	2 1/16	27	30-16d	4-½	2070	2760	1295	1725
MST37	12 ga	2 1/16	37	42-16d	6-½	2860	3815	1825	2435
MST48	12 ga	2 1/16	48	46-16d	8-½	3345	4460	2225	2970
MST60	10 ga	2 1/16	60	56-16d	10-½	4350	5800	2670	3565
MST72	10 ga	2 1/16	72	56-16d	10-½	4350	5800	2670	3565
PS218[4]	7 ga	2	18	—	4-⅝	—	—	—	—
PS418[4]	7 ga	4	18	—	4-⅝	—	—	—	—
PS720[4]	7 ga	7	20	—	8-½	—	—	—	—
HST2	7 ga	2½	21¼	—	6-⅝	—	—	3130	4175
HST5	7 ga	5	21¼	—	12-⅝	—	—	6385	8510
HST3	3 ga	3	25½	—	6-¾	—	—	4645	6195
HST6	3 ga	6	25½	—	12-¾	—	—	9350	12465

FLOOR-TO-FLOOR CLEAR SPAN TABLE

MODEL NO.	CLEAR SPAN	FASTENERS (TOTAL)	ALLOWABLE LOAD (133)
MST37	18	20-16d	1905
	16	22-16d	2100
MST48	18	32-16d	3135
	16	34-16d	3330
MST60	18	46-16d	4785
	16	48-16d	4990
MST72	18	56-16d	5800
	16	56-16d	5800
MSTI36	18	14-10dx1½	810
	16	16-10dx1½	930
MSTI48	18	26-10dx1½	1545
	16	28-10dx1½	1660
MSTI60	18	38-10dx1½	2330
	16	40-10dx1½	2455
MSTI72	18	50-10dx1½	3065
	16	52-10dx1½	3190

See CS for definition of clear span.

1. Maximum loads have been increased for wind or earthquake loading with no other increase allowed. Floor loads may not be increased for other load durations.
2. 16d sinkers (9 gauge x 3¼") or 10d commons may be substituted where 16d commons are specified at 0.83 of the table loads.
3. Allowable bolt loads are based on parallel-to-grain loading and the following minimum member thicknesses: MST–2½"; HST2 and HST5–4"; HST3 and HST6–4½".
4. PS strap design loads must be determined for each installation. Bolts are installed both perpendicular and parallel-to-grain.

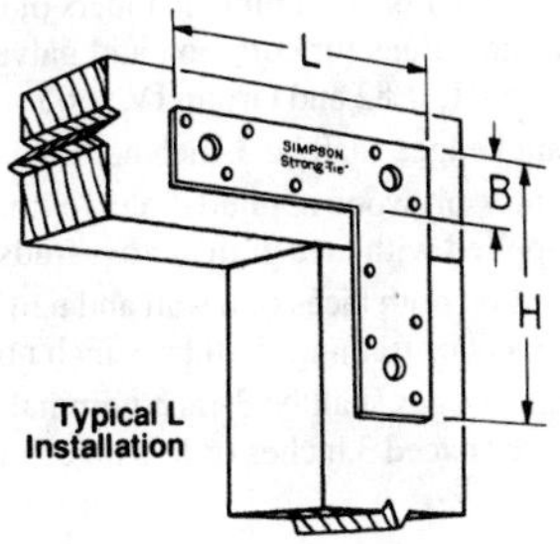
L
B
H
SIMPSON Strong-Tie

Typical L Installation

Reprinted with permission from *Connectors for Wood Construction*, Product & Information Manual © 1993, Simpson Strong-Tie Company, Inc.

T and L Strap Ties

MODEL NO.	MATL	DIMENSIONS			FASTENERS	
		L	H	B	NAILS	BOLTS
55L	16 ga galv	4¾	4¾	1¼	5-10d	—
66L	14 ga galv	6	6	1½	10-16d	3-⅜
88L	14 ga galv	8	8	2	12-16d	3-½
1212L	14 ga galv	12	12	2	14-16d	3-½
1212HL	7 ga ptd	12	12	2½	—	4-⅝
1616HL	7 ga ptd	16	16	2½	—	4-⅝
66T	14 ga galv	6	5	1½	8-16d	3-⅜
128T	14 ga galv	12	8	2	12-16d	3-½
1212T	14 ga galv	12	12	2	12-16d	3-½
1212HT	7 ga ptd	12	12	2½	—	6-⅝
1616HT	7 ga ptd	16	16	2½	—	6-⅝

Because of the versatility of these utility strap ties, load values are not given.

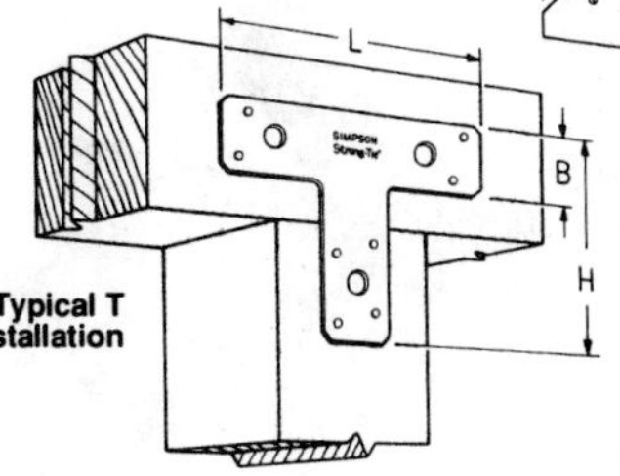
L
B
H
SIMPSON Strong-Tie

Typical T Installation

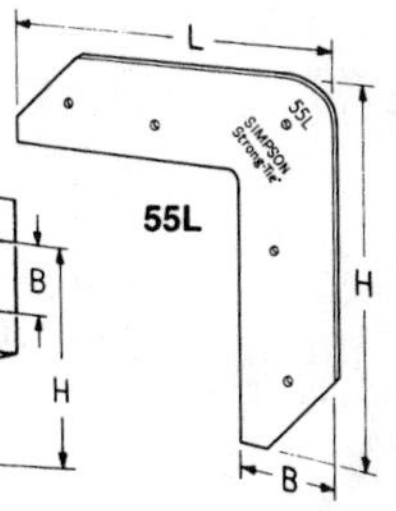
L
55L
SIMPSON Strong-Tie
H
B

Reprinted with permission from *Connectors for Wood Construction*, Product & Information Manual © 1993, Simpson Strong-Tie Company, Inc.

APPENDIX L
Allowable Shear for Wind or Seismic Forces in Pounds per Foot
for Plywood Walls with Framing of Douglas Fir-Larch or Southern Pine
[UBC Table 25-K-1]

PLYWOOD GRADE	MINIMUM NOMINAL PLYWOOD THICKNESS (Inches)	MINIMUM NAIL PENETRATION IN FRAMING (Inches)	NAIL SIZE (Common or Galvanized Box)	PLYWOOD APPLIED DIRECT TO FRAMING — Nail Spacing at Plywood Panel Edges				NAIL SIZE (Common or Galvanized Box)	PLYWOOD APPLIED OVER 1/2-INCH OR 5/8-INCH GYPSUM SHEATHING — Nail Spacing at Plywood Panel Edges			
				6	4	3	2[2]		6	4	3	2[2]
STRUCTURAL I	5/16	1 1/4	6d	200	300	390	510	8d	200	300	390	510
	3/8	1 1/2	8d	230[3]	360[3]	460[3]	610[3]	10d[5]	280	430	550	730[2]
	15/32	1 1/2	8d	280	430	550	730	10d[5]	280	430	550	730
	15/32	1 5/8	10d[5]	340	510	665	870	—	—	—	—	—
C-D, C-C STRUCTURAL II, plywood panel siding and other grades covered in U.B.C. Standard No. 25-9.	5/16	1 1/4	6d	180	270	350	450	8d	180	270	350	450
	3/8	1 1/4	6d	200	300	390	510	8d	200	300	390	510
	3/8	1 1/2	8d	220[3]	320[3]	410[3]	530[3]	10d[5]	260	380	490	640
	15/32	1 1/2	8d	260	380	490	640	10d[5]	260	380	490	640
	15/32	1 5/8	10d[5]	310	460	600	770	—	—	—	—	—
	19/32	1 5/8	10d[5]	340	510	665	870	—	—	—	—	—
			NAIL SIZE (Galvanized Casing)					NAIL SIZE (Galvanized Casing)				
Plywood panel siding in grades covered in U.B.C. Standard No. 25-9	5/16	1 1/4	6d	140	210	275	360	8d	140	210	275	360
	3/8	1 1/2	8d	130[3]	200[3]	260[3]	340[3]	10d[5]	160	240	310	410

[1]All panel edges backed with 2-inch nominal or wider framing. Plywood installed either horizontally or vertically. Space nails at 6 inches on center along intermediate framing members for 3/8-inch plywood installed with face grain parallel to studs spaced 24 inches on center and 12 inches on center for other conditions and plywood thicknesses. These values are for short-time loads due to wind or earthquake and must be reduced 25 percent for normal loading.

Allowable shear values for nails in framing members of other species set forth in Table No. 25-17-J of U.B.C. Standards shall be calculated for all grades by multiplying the values for common and galvanized box nails in STRUCTURAL I and galvanized casing nails in other grades by the following factors: Group III, 0.82 and Group IV, 0.65.

[2]Framing at adjoining panel edges shall be 3-inch nominal or wider and nails shall be staggered where nails are spaced 2 inches on center.

[3]The values for 3/8-inch-thick plywood applied direct to framing may be increased 20 percent, provided studs are spaced a maximum of 16 inches on center or plywood is applied with face grain across studs.

[4]Where plywood is applied on both faces of a wall and nail spacing is less than 6 inches on center on either side, panel joints shall be offset to fall on different framing members or framing shall be 3-inch nominal or thicker and nails on each side shall be staggered.

[5]Framing at adjoining panel edges shall be 3-inch nominal or wider and nails shall be staggered where 10d nails having penetration into framing of more than 1 5/8 inches are spaced 3 inches or less on center.

INDEX

Entries from footnotes are indicated by page numbers in italics.

INDEX OF FIGURES AND TABLES

PROFESSIONAL PUBLICATIONS, INC. ● Belmont, CA

INDEX BY UBC SECTION

Entries from footnotes are indicated by page numbers in italics.

INDEX OF PROBLEM TYPES

This index consists of a listing of specific elements that are calculated in the solved examples and practice problems in this book. This listing is provided to assist you in solving problems that may have similar parts and structures. The page references given refer to the pages on which the problem statements appear, not necessarily the pages on which the calculations are performed. Also, intermediate types of calculations are listed in addition to end results. Most entries have been listed several ways, so if you cannot find an item, try rephrasing it.

PROFESSIONAL PUBLICATIONS, INC. ● Belmont, CA